Orchids of the Northeast
A Field Guide

William K. Chapman
With a Foreword by Eric Lamont

Syracuse University Press

Copyright © 1997 by Syracuse University Press
Syracuse, New York 13244-5160

All Rights Reserved

First Edition 1997

97 98 99 00 01   6 5 4 3 2 1

Except when noted, all photographs are courtesy of the author.

The paper used in this publication meets the minimum requirements of American National Standard for Information Sciences—Permanence of Paper for Printed Library Materials, ANSI Z39.48-1984. ∞™

Library of Congress Cataloging-in-Publication Data

Chapman, William K., 1951–
    Orchids of the Northeast : a field guide / William K. Chapman ; with a foreword by Eric Lamont.
        p. cm.
    Includes bibliographical references (p.   ) and index.
    ISBN 0-8156-2697-5 (cloth : alk. paper). —ISBN 0-8156-0342-8 (pbk. : alk. paper)
       1. Orchids—New England—Identification.   2. Orchids—New York (State)—Identification.   I. Title.
QK495.064C355   1996
584'.15'0974—dc20   96-5733

Printed in Hong Kong

To Valerie, my wife and my best friend and companion, and to Andy and Carey, the two brightest flowers in our lives.

William K. Chapman, a biology teacher and member of the adjunct faculty at Utica College of Syracuse University, is the author of *Plants and Flowers: An Archival Sourcebook, Pheasants under Glass,* and individual field guides to the trees, mammals, and birds of the Adirondacks. His nature photography has appeared in *National Geographic* and the *New York State Conservationist.*

Contents

Colorplates ix

Foreword, Eric Lamont xiii

Acknowledgments xv

Introduction xvii

What Is an Orchid? 3

Key to Orchids of New York and New England 8

Group 1
Cypripedium 12
Calypso 33

Group 2
Calopogon 38
Arethusa 41
Pogonia 44

Group 3
Isotria 48

Group 4
Goodyera 54
Spiranthes 66

Group 5
Tipularia 88
Aplectrum 91

Corallorhiza 94

Triphora 105

Galearis 108

Amerorchis 110

Epipactis 113

Coeloglossum 116

Platanthera 118

Malaxis 163

Listera 171

Liparis 182

Glossary 189

Bibliography 193

Index 197

Colorplates

Cypripedium
arietinum   15
candidum   17
acaule   20
parviflorum var. *pubescens*   24
parviflorum var. *parviflorum*   26
parviflorum var. *makasin*   28
reginae   30
reginae forma *albolabium*   32

Calypso
bulbosa   33

Calopogon
tuberosus   38

Arethusa
bulbosa   41

Pogonia
ophioglossoides   44

Isotria
verticillata   49
medeoloides   51

Goodyera
tesselata   58
repens   60
pubescens   62
oblongifolia   64

Spiranthes
tuberosa   71
lacera var. *lacera*   73
vernalis   75
casei   77

lucida 79
ochroleuca 81
cernua 83
romanzoffiana 85

Tipularia
discolor 88

Aplectrum
hyemale 91

Corallorhiza
maculata 96
trifida 98
striata 100
odontorhiza 102

Triphora
trianthophora 105

Galearis
spectabilis 108

Amerorchis
rotundifolia 110

Epipactis
helleborine 113

Coeloglossum
viride 116

Platanthera
psycodes 123
grandiflora 125
leucophaea 127
lacera 129
blephariglottis 131
ciliaris 133
cristata 135
cristata, yellow form, Virginia 136
pallida 137
× *andrewsii* 139
hookeri 142

orbiculata 146
macrophylla 147
dilatata 151
hyperborea 153
huronensis 155
flava 157
clavellata 159
obtusata 161

Malaxis
brachypoda 165
unifolia 167
bayardii 169

Listera
australis 174
cordata 176
auriculata 178
convallarioides 180

Liparis
lilifolia 183
loeselii 185

Foreword

Eric Lamont
President, Long Island Botanical Society

THE UNEXPECTED DISCOVERY of a wild population of native orchids is one of the most rewarding experiences of a nature enthusiast. Botanical literature contains myriad accounts of such discoveries and encounters; often these experiences convey a deep emotional or spiritual quality. Upon finding *Calypso* for the first time, the great naturalist John Muir "shed tears of joy over its beauty."

Orchids are the sirens of the plant world; their beauty and biological mysteries lure inquisitive observers. I recall botanizing in the Great Smoky Mountains with the late Arthur Cronquist, the world-famous botanist from the New York Botanical Garden. We had been focusing our field studies on spring composites when Art located the lily-leaved twayblade in the rich mountain woods. He stopped in his tracks, sat down on the ground, and stated, "The day I am unimpressed by a wild orchid is the day that I stop studying plants." Orchids excite our subjective emotions and our objective quest to understand the world.

Early botanists were inspired by contemplating orchids, and they conjured up images of fiery-red dragons, celestial sheep, swarms of winged insects, marine corals, gargoyles, and beautiful nymphs and goddesses from Greek mythology. Scientifically inclined investigators also have been captivated by the complexities of the orchid flower. After publishing his landmark book on evolution, Charles Darwin devoted years to studying "various contrivances by which orchids are fertilized by insects."

William Chapman has artfully intertwined our scientific understanding of orchids with romanticized accounts and historical background. Detailed and concise descriptions of species are presented in a reader-friendly but scientifically accurate style. Current research on taxonomically difficult groups of species is explained and interpreted.

The northeastern States represent less than 5% of the land mass of continental United States where more than 125 native species of orchids occur north of Florida; yet 60 orchid species are known to occur in the northeast. This relatively rich orchid flora reflects a diversity of habitats where several orchids reach their northern or southern limits.

Each native species of orchid in the northeast has experienced a decline in population size during the past 100 years. The native orchids of Long Island, New York, have suffered most heavily from human population growth. Historically, 37 orchid species have been known to occur on Long Island; however, 10 of the species have not been observed during the past 60 years and three of them have not been reported in over 100 years. Throughout the northeast several extant populations of rare orchid species are on the verge of extirpation because of human activity.

As we rapidly approach the close of this century, it is a time to take stock of our natural heritage. The present review of our exquisitely beautiful and intriguing orchids should do much to encourage our enthusiasm to preserve them and their native habitats. This hope for the future is the ultimate goal depicted in this thought-provoking volume, to encourage the reader to "appreciate both the plants and how fragile and endangered their existence is in our increasingly overdeveloped environment."

Acknowledgments

MANY PEOPLE CONTRIBUTED valuable time and effort on behalf of this book. It is to Charles J. Sheviak that I owe the greatest debt. His expertise, his advice, and his unselfish sharing of his time through innumerable phone conversations, letters and meetings were invaluable in guiding my efforts. John Freudenstein, Michael Homoya, Eric Lamont and Charles Sheviak all reviewed the text and offered important perspectives on improvising the contents. Eric Lamont also kindly agreed to provide the Foreword. Alan Bessette reviewed the final text and made valuable technical suggestions. Paul Martin Brown freely shared his time and experience during numerous photographic field trips across New York and New England. Evelyn Greene kindly guided me to several important Adirondack habitats.

The following people also all provided many forms of assistance ranging from advice to introductions to invitations onto their property. If I have inadvertently omitted anyone who deserves to be on this list, I sincerely apologize. My thanks to George Beatty, Doug Bassett, Debbie Benjamin, Philippa Brown, Frederick Case, Jr., Steven Daniel, Robert Dirig, Lee Drake, Cordelia Flannigan, Nancy Hinman, Charles Johnson, Fay Lyon, Anne McGrath, Everett Marshall, Betsy Mercelis, Ed Miller, Richard Mitchell, John Page, Oakes Plimpton, Allan Reddoch, Joyce Reddoch, Kathy Regan, Mona Rynearson, Janet Spring, Russell Spring, F. Robert Wesley, and Stephen Young. I am also especially grateful to Dr. Robert Mandel and his staff at Syracuse University Press for making this book possible.

Introduction

ORCHIDS! The name of no other group of plants arouses such immediate interest and attention. No other plants invoke such a strong desire to know more about them, and certainly no group of plants is in such desperate need of human understanding and compassion, for in the northeast far too many of their kind stand poised on that brink from which there is no return.

The orchid bug bit me early in life, and I have been blessed by being able to pursue that study here in New York and New England, in the southeast, in the American Rocky Mountains, in eastern and western Canada, and in Hawaii. Hawaii came as something of a surprise to me. In 1986 and 1987 I was fortunate enough to pursue intensive natural and cultural researches in that state, and my early expectations were that myriad species could be found in Hawaii's diverse habitats. In reality I found only six, and the three commonest of these were Asian introductions! The realization that there were approximately ten times as many orchid species in the northeastern United States set the stage for the creation of this volume. Within this region a number of important northern species reach their southern range limit, and a number of equally fascinating southern species are found at their northernmost boundaries. Except for the cold fact of winter weather limiting our orchid observations to the warmer months, the northeast may well be one of the most interesting areas in which to pursue orchid studies in the United States.

Interest in orchids is nothing new. In the history of botanical literature, few if any other families of plants have had so many volumes penned and published solely on their respective members. From around the turn of this century through the 1930s, most orchid texts were written for the casual reader and amateur naturalists. In the latter half of this century, that trend was reversed, with many if not most texts being written in formal botanical language that more exactly defined the true nature of each species. The downside of such studies is that, even though they more accurately reflect our ever increasing scientific understanding, they do so in a terminology that is largely inaccessible to the average person.

This book is an attempt to bridge those two divergent styles, to take the best of each and to present it in everyday language. It will not attempt to compete with the best of the scientific works. Correll in 1950 and Luer in 1975 have already produced classics in that vein. This book does present the latest information available on orchid identifications and classifications, but it attempts to do so in a reader-friendly style. The Comments that follow the Descriptions are, however, somewhat more relaxed in character. They are in many cases very intentional throwbacks to those early amateur-oriented orchid books, when plants had personalities and hard botanical facts were interpreted and presented in a romanticized manner.

An important set of measurements included in the descriptions is the overall height and width dimension for each species. In this work these measurements are limited to the sepals and the petals, which are collectively referred to as the *perianth*. The measurements do not include the spurs or ovaries, which are technically parts of the flower but which are not used here because of their frequently unusual shapes and orientations. Please keep in mind that since these measurements include both sepals and petals, some species such as the whorled pogonias and the lady's slippers with long and widely spreading sepals may generate some surprisingly large dimensions. In a few cases, such as *Malaxis,* it is typical to find flowers in many orientations on each plant. For those few species the measurements are given for specific parts of the flower, such as from the tip of the lip to the tip of the dorsal sepal. Speaking of measurements, the dimensions given for floral parts in the descriptions are those typical for an adult plant growing in the optimum habitat for that species. In the field you may encounter both smaller and occasionally larger specimens.

Many of the liberties just discussed are made possible by the fact that this is a regional work limited to only one corner of the United States. Large and complex groups such as the ladies' tresses have only a small number of species in the northeast. With fewer species and fewer look-alikes, there is less need to rely on a greater number of physically minute traits. Another advantage of working within a limited geographic range is that it is possible to be far more precise about flowering seasons than can be done in a continental monograph. A good example of this is *Spiranthes vernalis*. Its range extends from Central America to southeastern Canada. Accurate descriptions of this species correctly state that it blooms from February (in the south, hence its common name of the "spring ladies' tresses") to August (its season in the northeast, where an equally accurate local common name could have been the "late summer ladies' tresses").

This work will present only flowering seasons that are accurate for the northeast. Since the northeast is in itself large enough to generate a range of bloom dates from different localities, this book will present two sets of flowering seasons. The first of these will be the range within which 95 percent of the orchids for any species bloom within the entire northeast. These dates were collected by examining voucher specimens in herbarium collections. The second and more exacting set of dates will be taken from a mythical multihabitat area near Piseco Lake in the Adirondack Mountains, where most northeastern species of orchid may be found growing within close proximity to each other. These dates, which are given in parentheses, can be used to demonstrate the order of flowering that occurs between closely related species in most (there will always exist exceptions to every generalization!) habitats and situations. For the few species that live in only one small portion or corner of the northeast, such as the previously mentioned *Spiranthes vernalis,* only dates for the area they actually inhabit will be listed.

The study of orchids can become one of the most fascinating aspects of botany available to naturalists at any level of expertise. It is my hope that the limited information presented here encourages the reader to go on and learn far more, and in the process come to appreciate both the plants and how fragile and endangered their existence is in our increasingly overdeveloped environment. While many species of orchids still exist in large viable populations in the northeast, *Amerorchis, Aplectrum, Calypso,* and others have suffered major declines. What a wonderful thing it would be if these orchids could become the green equivalent of the bald eagle and the buffalo, that we would take the necessary steps to halt the decline of their kind and instead initiate greater efforts aimed at protecting and increasing their numbers. Through the beauty of their physical forms and their enchanting natural histories and habits, the orchids have added an element of magic to our lives. Can we give them the gift of survival in return?

William K. Chapman

July 1995
Utica, New York

Orchids of the Northeast

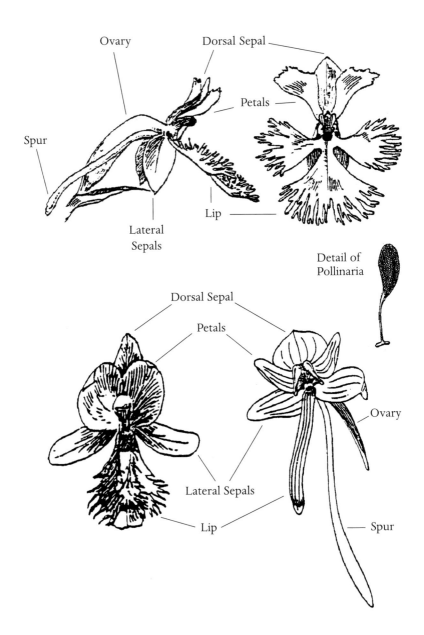

What Is an Orchid?

WHAT IS AN ORCHID? Where does it fit into the classic classification schemes that encompass all living things? What we need is to identify that combination of characteristics that will allow us all to act as amateur Sherlock Holmeses and deduce the possible identity of any unfamiliar orchid we happen to encounter. Gibson (1905) penned the following tribute: "The orchids are the nobility of the flower world. They are a race set apart. No common garb is theirs. Like scions of a noble race, they have a curiously devised heraldry and may be known by their horns and antlers, tails and ears, and queerly fashioned crests." What we will do here is to examine those unique vegetative "horns and antlers, tails and ears" that set the orchids apart from all other plants.

When we examine any flower the first fact we observe is the obvious, that it is indeed a flower. This fact places the bloom among angiosperms (flowering plants) and automatically assumes that it must belong to one of the two great angiosperm divisions, the dicotyledons or the monocotyledons, also known as dicots or monocots for short. The terms *dicotyledon* and *monocotyledon* mean "two seed-leaves" and "one seed-leaf" respectively, but as this refers only to a plant's first days of life they are not useful tools in identifying a mature plant. Fortunately, there are a pair of associated traits which can be easily and safely (not requiring a damaging dissection) observed on an adult plant. Dicots exhibit a complex branching pattern on the leaves, and their floral parts are usually arranged in multiples of two, four, or five. Monocots typically exhibit parallel veining on the leaves, and have floral parts arranged in threes. These two traits are usually sufficient (except for a few orchids that lack leaves at flowering time) to correctly assign those orchids we encounter to the monocot grouping.

Inclusion in the monocots finds our orchids related to such well-known plant families as the lilies and the irises, and we will now turn to the floral parts to realize how truly unique the orchids are. We have already stated that the monocots have floral parts in threes. In most monocots (orchids and grasses excluded) the floral parts are very symmetrically arranged. In such well-known flowers as trilliums, we observe three large, nearly identical petals, and directly behind these we find three prominent sepals. In irises and many ornamental lilies, the sepals may appear as large and brightly colored as the true petals, while in other species they are smaller and less colorful, but in any case they are symmetrically arranged.

Like the lilies the orchids are six-parted (three petaled and three sepaled), but here the similarity ends, for orchids are very irregular in appearance. If we visually bisect an orchid flower vertically, we find that the lateral sepals form nearly identical mirror-images of each other. The third sepal, which usually is the uppermost floral part on an orchid flower, is sometimes identical to the other two, but more often it is larger than the side sepals and frequently grows forward, providing a sheltering hood-like structure over the inner reproductive parts.

Like the lateral sepals, the lateral petals are usually nearly identical, and frequently join with the dorsal sepal in sheltering the orchid's delicate center. It is the third central petal, typically referred to as the *lip* (for it does usually descend downwards from the "face" of these flowers), for which the orchids are best known. As we shall soon see, the lips of the various species have evolved into an amazing variety of sizes, shapes, colors and presumedly smells and tastes, all for the purpose of luring an equally varied array of insects into the flower to assist in the process of fertilization. In most orchids, the lip secretes a nectar on which the pollinating insect feeds. The orchids demonstrate an amazing number of evolutionary strategies to match these liquid lures to the exact size and shape of the specific insect species that assists in the plant's reproduction. Since sex is sometimes a greater lure than food, some insects are enticed to enter orchids that seem to mimic their natural mates, even though the only reproduction then occurring is that which benefits the floral trickster. Apparently this orchid learned the secrets of sex appeal advertising long before Madison Avenue did! Depending on the species, the lip may produce the nectar near its base, inside a large inflated pouch, or at the tip of a long hollow tube called a *spur,* also referred to as a *nectary* earlier in this century.

In most species of monocots this central petal is positioned as the uppermost floral part. The upper vertical orientation is referred to as *nonresupinate.* In most but not all orchids this orientation is reversed, with the petal-lip being the lowest floral part, a condition in the flower known as *resupinate.* In a resupinate flower the entire bloom has been tipped upside down. One possible reproductive advantage to this positioning is that it allows the oversized and frequently irregularly shaped lip to serve as a landing pad or runway on which airborne insects can alight and obtain a solid foothold from which to explore the visual and aromatic beacons which guided it to this flower.

We may wonder what is responsible for twisting the flower upside down and also wonder how this is carried out. In orchids the culprit is the combined ovary-pedicel, often erroneously called just the *ovary* in many orchid descriptions. Technically, the pedicel is the stalk of an individual flower, and the ovary is the basal portion of the female floral parts where seed development takes place. An orchid has what is referred to as an *inferior ovary,* an ovary that occurs beneath the floral parts. Because both the pedicel and the ovary are found between the main stem (from which many pedicels may spring) and the brightly colored sepals and petals, these two parts are often visualized as a single, continuous part. In monocots such as lilies the ovary is perfectly straight, but in most orchids the ovary twists completely around as it develops, resulting in the upside down/lip lower resupinate position of the orchid flower.

Another interesting aspect of orchid ovaries is that they are often unusually large in proportion to the floral parts. Among some smaller orchids such as the *Corallorhizas* and many of the *Platantheras* this creates the illusion of small delicate blooms growing from fat, stout "stalks" as large or larger than the flowers themselves. Despite the disproportionately large size of the ovaries, the seeds that are matured within are minute. The size of an individual seed released from a dried and matured ovary (called a *capsule*) may be smaller than a period on this page! One reason orchid seeds are so small is that they lack the supply of stored food that makes up the bulk of most plant seeds, including such popular foods as corn, wheat, and other breadstuffs. Because nature has not provided the developing orchid embryo with a sustaining supply of food to carry it through the first days of life, it has a precariously short time to obtain an alternate source of nutrients. Research and observation have revealed that the orchid seedlings survive by forming a symbiotic (mutually beneficial) mycorrhizal relationship by intertwining its roots with the subterranean rootlike mycelia of a fungus. Presumedly, each member of this partnership contributes needed nutrients to the other, and for the orchid at least no survival is possible without this sharing relationship. Not many orchid seeds succeed in establishing this relationship, so it is fortunate that the average orchid capsule produces so many seeds, well in excess of one million in some species! It is clear that what orchid seeds lack in size they make up for in astronomically large numbers.

If we return to the center of our "typical" orchid flower, we encounter what many people consider to be the orchid's most unique reproductive adaptation, the column. In most plant families the male and female parts of each flower are separate, in some cases male and female parts occurring in separate blooms or even on separate plants. In orchids the opposite is true, with the male stamens and their anthers fusing to the female stigma and style, forming a single reproductive organ. This growing together of the reproductive organs is the one feature that distinguishes the orchids from all other flowers. The structure of the column can vary tremendously, and it is the premier characteristic used in assigning the various orchids to their specialized genera. In the *Listera* the column may appear as a simple cylindrical structure in the center of the flower, while in the lady's slippers it may be so broad that it appears as an additional miniature petal.

In all angiosperms the anther is the male portion of the flower that provides the pollen. The number of fertile anthers on an orchid column is used to separate all orchids into two broad classifications, the diandrous and the monandrous orchids. Diandrous orchids have two fertile anthers, one on each side of the column. These include the best known of our wild orchids, the lady's slippers. In addition to possessing two fertile anthers, lady's slipper orchids also exhibit a staminode, a broad petal-like growth that gives their columns an almost fer-de-lance shape or outline. In contrast, monandrous orchids have only one fertile anther. In the northeast, this exceptionally large group includes all of the orchids except for the previously mentioned lady's slippers. At quick glance these single anther orchids are deceiving because that single anther usually sports two separate pollen masses. It is these dual pollen masses that are responsible for so many of our orchids being regarded as "face flowers" because to many imaginative naturalists they appear as two small vegetative "eyes" that peer back at us from within the blooms we examine. This is especially true of *Platanthera* orchids, for in addition to these pollen eyes the flowers possess a longitudinal bilateral symmetry (as do all animal faces), sometimes a conspicuous spur opening to suggest a mouth, and an ornate lip to serve as a chin, jaw, or even a heavily fringed beard.

Orchid pollen is packaged quite differently than that of most other flowers. Instead of consisting of unconnected powdery grains, orchid pollen is most often collected into small, sac-like structures at the tip of a short stalk. Each of these pollen packets is called a *pollinium* (pl. *pollinia*). At the base of each stalk is a sticky "foot," which fixes itself to the face or other body parts of the visiting pollinating insect so that all the pollen in a single bloom may be withdrawn from the flower at the same moment during the pollination process.

Naturalists have long chuckled at illustrations of bees leaving orchids with a long pollen mass glued to each of its compound eyes, giving the insect an instantly "horned" appearance. The unique physical structure and shape of each orchid has evolved to ensure that this act takes place, and expanded information on how selected species arrange their own variations of this reproductive drama will be found in the individual species descriptions.

Key to the Orchids of New York and New England*

To use this key, read through the five primary choices (I, II, III, IV, V) and select the one best describing the species you wish to identify. Within these primary choices, make your next choice from A, B, C, etc. Continue to narrow down your choices until the specimen is identified.

I. Flowers solitary or double on a single stem. Lip saccate, inflated, showy.

II. Flowers solitary to several (3–15 or more) on a single stem. Lip and rest of flower usually some shade of pink, showy, lip not inflated, ornamented with a prominent tuft of white to yellow "hairs." Leaves usually solitary.

  A. Flowers several (3–15 or more) on a single stem. Lip narrow, growing uppermost in the flower. *Calopogon tuberosus* → page 38.

  B. Flowers usually single on the stem. Lip wide, positioned on the lower portion on the flower.

III. Flowers usually solitary or sometimes double on a single stem. Leaves 5–6 and whorled about the stem just below the flower. *Isotria*, the whorled pogonias → page 48.

IV. Flowers several to many on a single stem, small and white, either somewhat spherical or tubular when viewed from the side.

  A. Flowers small, 3–9 mm (1/8–1/3") long, somewhat spherical when viewed from the side; leaves forming a rosette at the base of the stem, usually marked with a beautiful whitish reticulation. *Goodyera*, the rattlesnake plantains → page 56.

  B. Flowers small, 3–12 mm (1/8–2/5") long, tubular when viewed from the side; leaves broad, slender, or absent. *Spiranthes*, the ladies' tresses → page 69.

V. Flowers small to medium-sized, several to many on a single stem, but neither spherical or tubular when viewed from the side.

A. Leaves typically absent or dying at flowering time.

    B. Leaves typically present at flowering time. Lip may or may not have a prominent spur → page 104.

        1. Lip with a prominent spur. *Tipularia discolor* → page 88.

        2. Lip without a prominent spur, with a basal leaf that is typically shriveled at flowering. *Aplectrum hyemale* → page 91.

        3. Lip without a prominent spur; leaves always absent. *Corallorhiza*, the coral-roots → page 95.

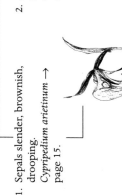

1. Lip margin wavy, uneven. Leaf long and slender, basal. *Arethusa bulbosa* → page 41.

2. Lip margin heavily fringed. Leaf lance-shaped, positioned near middle of stem. *Pogonia ophioglossoides* → page 44.

B. Lateral sepals not united.

    2. Sepals narrowly lance-like, rose-pink, erect, and spreading above the lip. *Calypso bulbosa* → page 33.

A. Lateral sepals united behind the lip to form a single pendent blade. *Cypripedium*, the lady's slippers → page 14.

    1. Sepals slender, brownish, drooping. *Cypripedium arietinum* → page 15.

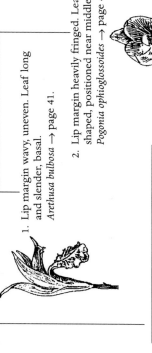

*Key based primarily on flowering characteristics.

Group 1

Cypripedium
Calypso

Cypripedium
The Lady's Slipper Orchids

Etymology: *cypri* = Kypris, an early Greek name of Venus, later Aphrodite, the Greek and Roman goddesses of love; *pedium* = "foot," "shoe," or "slipper."

Speak the word *orchid* to any person on the street, and the first flower they envision will likely come from the subfamily *Cypripedioideae* because the colorful lady's slipper-type flowers of domestication come from this group. Probably the plant structure most recognizable to the average person is the large and colorful inflated lip, a hallmark of this family. Sharp-eyed scientists who look a little closer will see the unique, double fertile anthers, which set this subfamily apart from all other orchids mentioned in this book. There are over one hundred species known from the four genera that make up the subfamily *Cypripedioideae*. The genus *Paphiopedilum* has supplied most of the cultivated lady's slipper-type flowers just mentioned, but we shall turn our attention for the moment to the *Cypripedium*, the native lady's slippers that can only be appreciated in their natural habitats.

Of the approximately thirty *Cypripedium* species that inhabit the temperate to cold portions of the northern hemisphere, eleven are found in the United States and Canada. Five of these can be found in the area covered by this book. Members of this lady's slipper group have heavily veined leaves, which may grow from the base of the plant or be arranged alternately up the stem. At the top of each stem is found a large leaf-like structure known as the floral bract. From the base of each bract springs a slipper-like flower. Floral bracts may be observed on other orchids, but in the lady's slippers their unusually large size and typically erect nature (in all species except the pink lady's slipper) form a beautiful green backdrop against which to view the flower. Some people have minimized the attractiveness of the sepals and petals surrounding the enlarged lip, noting that only in the lady's slipper of the Queen are they a bright (in this case white) color. However, I see their yellowish greens and purplish browns as the perfect framing counterparts to the showier lip.

Without a doubt, it is the large and colorful lip that draws our attention to these flowers. These same lips also attract the interest of bees and other insects, which then become tricked into assisting the pollination process. Lured by the bright colors and other factors, insects, usually bees, enter the inflated lip, only to find that the physical structure may not permit them to exit in the same way they entered. Instead, the bees spy either of a pair of much narrower "back doors" near the base of the lip. In squeezing through these openings the bee is forced to brush up against one of two fertile anthers, and the bee becomes smeared with sticky pollen grains. The bee finally frees itself from this floral trap and, apparently not one to learn from experience, promptly carries its load of hitchhiking pollen to the next slipper it encounters.

The bee entering the next flower has the pollen scraped off onto the waiting stigma by an ingenious but very logical structure before it picks up another load of pollen on its way out of this cleverly designed floral maze.

We can only conjecture about what bees see in these flowers, but our own species sees them through very imaginative eyes. More than any other flowers, orchids in general, and lady's slippers in particular, seem to inspire romantic flights of imagination in even the most straightlaced and stuffy scientists. Freud would be pleased to see the number of sexual symbolisms projected onto the flowers, which in many minds may represent either male or female genitalia, depending on which species is being scrutinized. To my mind the most romantic image assigned to a lady's slipper was provided by Neltje Blanchan (1916), who described the purplish pink markings on the lady's slipper of the Queen as "wine appearing to overflow the large white cup and trickle down its side." Blanchan goes on to describe the rare white form in which "sometimes unstained, pure white chalices are found."

The famous botanist Linnaeus applied the "shoe" or "slipper" name to the European yellow lady's slipper, but it seems likely that he was merely making official a common name that predated his contributions. Although the flower was originally named for Venus, the ancient goddess of love, the Christian church reworked this fantasy into the "slipper of Our Lady [the Virgin Mary]," which was later corrupted or shortened into the name of lady's slipper (Niles 1904) and (Morris and Eames 1929). Today this floral footwear appears in a number of styles and colors, from a petite white to yellow or the royal pink and white of the Queen. Because of its unique shape, another lady's slipper is seen as a sheep, a ram with his head down in the charging position. Put them all together (the "shoes" and "sheep" and "the Queen"), and we find ourselves visualizing a bizarre floral parade. If we place the participants of this parade in the order of their blooming, the ram's head lady's slipper would probably head the column of marchers. It is only an educated guess whether this or the white lady's slipper blooms first, since I know of no habitat where the two coexist. Happily, there is a location where the white, pink, yellow, and Queen's lady's slippers are all found in close proximity, and that sequence is the order in which the first blooms of each species appear. It seems only fitting that all other species precede the Queen, almost as if they are intentionally preparing the way for the largest and tallest northeastern orchid.

Key to Cypripedium: Lady's Slipper Orchids
   (continued from page 9)

I. Plant with two basal leaves. Flower solitary, pink or sometimes whitish, lip pendent.
   C. acaule → page 20.
II. Plant with alternate leaves on an upright stem.
   A. Lip white, sometimes with purplish markings inside, about 2.25 cm ($9/10''$) long.
      C. candidum → page 17.
   B. Lip white with pink, red, or purplish markings on the outside.
      1. Lip small, 1.5–2.5 cm ($3/5$–$1''$) long. Flowers usually solitary on the stem.
         C. arietinum → page 15.
      2. Lip large, 2.5–5 cm (1–2'') long. Flowers frequently in pairs.
         C. reginae → page 30.
   C. Lip yellow, 15–54 mm ($6/10$–$2 1/10''$) long.
      C. parviflorum → page 23.

Cypripedium arietinum R. Brown

Etymology: *arietinum* = "ram-like," as in Aries, the celestial sheep of the zodiac.

Common name: ram's head lady's slipper.

Description: Leaves 3–4, 5–10 cm (2–4″) long, lance-shaped, strongly veined, arranged alternately on the stem, bluish green to dark green. **Stem** 10–30 cm (4–12″) tall, pubescent, green, supporting 1 or rarely 2 flowers at the summit. **Flower** small, up to 3–3.5 cm (1⅕–1⅖″) tall, 2.5 cm (1″) wide, 2 cm (1⅘″) long, with a curiously shaped purple and white pouch-like lip, fragrant. **Sepals** and **petals** colored alike with a mixture of purplish brown and greenish streaking; dorsal sepal 1.5–2.5 cm (⅗–1″) long, broadly lance-shaped, angled over the lip at about 45° and drooping lower following pollination; lateral sepals and petals similar, 1.5–2 cm (⅗–¾″) long, slender, the sepals slightly shorter and broader, all 4 drooping below the lip; **lip** 1.5–2 cm (⅗–⅘″) long, an elongated pouch open at the top and with a wide, blunt, "spur-like growth" oriented downward from the tip, around the opening the lip is white and coated with conspicuous white hairs, the sides are white with pale purplish or reddish netting, the spur-like growth is a darker purple or red.

Habitat: on rocky soil under white cedar *(Thuja occidentalis)* or creeping juniper *(Juniperus communis),* usually along lakeshores, or in cool sphagnum fens.

Flowering season: third week of May to second week of June (third or fourth week in May).

Range: from Maine to Massachusetts west around the Great Lakes and Lakes Winnipeg and Manitoba. In the northeast it is found in central New York and the Adirondack Mountains, western Massachusetts, Vermont, most of New Hampshire and the southern two-thirds of Maine. In many parts of this range its distribution is very spotty, with only a few scattered populations.

Comments: The ram's head is structurally quite unlike our other lady's slippers. It is the only species whose lateral sepals are not joined or united behind the lip. It is also the only species with the curious spur-like structure on the lower surface of the lip. This "spur" is responsible for the common name of ram's head, appearing as the lowered face of a charging ram, leaving the upper more inflated parts to represent the horns and shoulders of our combative "slipper-sheep."

Cypripedium candidum Muhlenburg ex Willdenow

Etymology: *candidum* = "bright white," referring to the lip.

Common names: small white lady's slipper, white lady's slipper.

Description: Leaves 3–4, 7.5–18 cm (3–7″) long, lance-shaped with the bases sheathing the stem, strongly veined, arranged alternately on the stem, green. **Stem** 15–35 cm (6–14″) tall, somewhat pubescent, green, supporting 1 or rarely 2 flowers at the summit. **Flower** about 4–5 cm (1³⁄₅–2″) high and wide, 2.5 cm (1″) long, with a white pouch-like lip. Dorsal **sepal** about 2.5 cm (1″) long, lance-shaped, angled upward at about 45° over the lip, with a wavy margin, pale green with brownish or purplish margins and veins; lateral sepals united to form a single lance-shaped blade positioned behind the lip, slightly smaller than the dorsal sepal but similarly colored; **petals** about 3 cm (1¹⁄₅″) long, slender, twisted, spreading and slightly drooping, colored like the sepals; **lip** about 2.25 cm (⁹⁄₁₀″) long, an elongated pouch open at the top, waxy white outside, purple-veined inside.

Habitat: moist marl meadows, often "tucked away" under sheltering shrubs.

Flowering season: third week of May to first week of June. In New York this orchid is almost always in bloom the last week in May. The species flowers for about two weeks, with the first blooms appearing in mid to late May, dependent on the severity of the springtime floods over the marl meadows.

Range: very local from southern Manitoba through the upper Mississippi River drainage and eastward to New York and Pennsylvania. In the northeast the white lady's slipper is known only from a few isolated specimens in New York.

Comments: Morris and Eames (1929) provided the best brief description of this species by comparing the lip to the "pearly purity . . . [of] a swallow's egg." The best known and most accessible population of the white lady's slipper in the northeast occurs in a marl meadow owned by a major conservation organization. This situation is not as ideal for the orchid as it first sounds, for this small area has been penetrated by the familiar alien invader, *Phragmites*. Left unchecked, these tall reeds could overrun the meadow and choke out the already endangered native orchids. A solution that could control the spread of these reeds might be to attack the *Phragmites* by mechanical means and thus help the lady's slippers survive. Currently, this area is being studied to develop a plan to manage it in a manner that will hopefully maintain this endangered species.

Sharing the same aforementioned marl meadow with the small white lady's slipper and the large phragmites is the potentially deadly massasauga rattlesnake, which on occasion has introduced an element of risk into the normally tame "sport" of orchid photography.

In the western portions of its range, the white lady's slipper has been known to hybridize with both the large and small varieties of the yellow lady's slipper. In both of these crosses the offspring are midway between their parents in terms of both size and color. *Cypripedium* × *andrewsii* nm. *andrewsii*, a cross between *C. candidum* and *C. parviflorum* var. *makasin*, has dark sepals and petals and an off-white to pale yellow lip with reddish purple markings near the openings. *Cypripedium* × *andrewsii* nm. *favillianum*, a cross between *C. candidum* and *C. parviflorum* var. *pubescens*, has pale green sepals and petals with brownish tips and margins and an off-white to pale yellow lip. I am not aware of either hybrid occurring in the northeast, but I do know of a location in New York where both species bloom at the same time and may be found in close proximity, so the possibility of such populations cannot be discounted.

Cypripedium acaule Aiton

Etymology: *acuale* = "stemless." This rather confusing description arises from an attempt to separate the leafless above-ground stem of this species from the leaf-bearing stems of other lady's slippers.

Common names: moccasin flower, pink moccasin flower, pink lady's slipper, stemless lady's slipper.

Description: Leaves 2, 10–22 cm (4–8¾") long, broadly lance-shaped, strongly veined, basal, pubescent, usually rich green above, silvery green to yellow-green on lower surface. **Stem** 15–30 up to 38 cm (6–12 up to 15") tall, pubescent, supporting a single flower at the summit. **Flower** large and showy, about 7 cm or more (2¾" or more) tall and wide, 3 cm (1⅕") long, with a pink pouch-like lip, pendent. **Sepals** and **petals** usually light purplish brown but may be yellowish green to purplish; dorsal sepal 4.5–5 cm (1¾–2") long, lance-shaped, usually angled forward as if to shelter the lip, lateral sepals united to form one lance-shaped blade positioned behind the lip, slightly smaller than the dorsal sepal; petals 4–6 cm (1⅗–2⅖") long, narrower than the sepals, often loosely twisted, drooping; **lip** 4–6 cm (1⅗–2⅖") long, an egg-shaped pouch with a vertical opening down the front, margins of the opening deeply enrolled, pale pink to pink with darker pinkish-red veinings, much paler on the first day of flowering.

Habitat: highly variable, from dry coniferous or mixed forests to sphagnum fens and swamps.

Flowering season: third week of May to second week of July (fourth week in May to fourth week in June).

Range: from Newfoundland west to Great Bear Lake in Canada's Northwest Territories, south to the upper parts of Georgia and Alabama. It is found throughout the northeast.

Comments: Within much of its range, the moccasin flower is the most commonly encountered lady's slipper. With the probable exception of the helleborine orchid in the northeast, it is also the most frequently encountered and easiest to identify woodland orchid of any species, making it a favorite of casual hikers. The common name of moccasin flower is a New World interpretation of the Old World idea that the inflated lip of the lady's slippers suggests fancy footwear. Albino forms, with pure white or nearly white lips, occur occasionally in the northeast.

The Yellow Lady's Slippers

At different times in botanical history the yellow lady's slippers have proven to be quite a taxonomic challenge. Classification schemes have been proposed which elevated most of the major varieties to species status. Others have suggested that as many as ten distinct varieties of one highly variable species exist. Compounding the confusion is the fact that the physical form of an individual orchid can be influenced by whether it inhabits an ideal or a stress-inducing habitat. Until recently, most studies accepted the belief that the North American yellow lady's slippers were varieties of the superficially similar European yellow lady's slipper, *Cypripedium calceolus*. Charles Sheviak (1994, 1995) has presented evidence that the North American plants represent at least three varieties of a distinct American species, *C. parviflorum*. Sheviak has, however, expressed the belief that this interpretation of the yellow lady's slippers still contains a number of potential problems and that these classifications may need to be amended in the future.

If I were to compare the previous classifications with Sheviak's recent modifications, I would suggest that *C. calceolus* var. *pubescens* and var. *planipetalum* (found just north of the area covered by this work) are now transferred to *C. parviflorum* var. *pubescens*, while the older *C. calceolus* var. *parviflorum* descriptions included both *C. parviflorum* var. *parviflorum* and var. *makasin*. In fairness I must point out that this comparison between the two classifications is my own and that, while I believe it will hold up for the great majority of specimens examined, some exceptions probably occur. The key below is a rewording of Sheviak's key, written in a style intended to be more accessible to the average reader. In the descriptions that follow the key, I have attempted to combine Sheviak's key with other physical traits and aspects of natural history, such as flowering seasons. If future investigations reveal errors within the descriptions, they should be regarded as my own.

Key to the Varieties of Cypripedium parviflorum
(continued from page 14)

I. • Lip large, up to 54 mm (2$^{1}/_{10}$″) long. Sepals and petals usually pale, unmarked or with varying amounts of reddish brown to madder markings.
 • Outer surface of tubular bract below the leaves densely and silvery pubescent when young.
 • Fragrance moderate to faint, rose-like or pungent-musty.
 var. *pubescens* → page 24.

II. Lip small, 15–34 ($^{6}/_{10}$–1$^{3}/_{10}$″ long). Sepals and petals appear uniformly dark reddish brown.
 A. • Lip 22–34 mm ($^{9}/_{10}$–1$^{3}/_{10}$″ long).
  • Fragrance moderate to faint, rose-like or pungent-musty.
  • Color of sepals and petals a result of being densely and minutely spotted with reddish brown.
  • Outer surface of tubular bract below the leaves densely and silvery pubescent when young.
  var. *parviflorum* → page 26.
 B. • Lip 15–29 mm ($^{6}/_{10}$–1$^{1}/_{10}$″) long.
  • Fragrance intense, sweet.
  • Sepals and petals uniformly suffused with reddish brown (some patterning may be visible near the base).
  • Outer surface of tubular bract below the leaves inconspicuously pubescent or smooth when young.
  var. *makasin* → page 28.

Cypripedium parviflorum Salisbury var. **pubescens** (Willdenow) knight

Etymology: *parviflorum: parvi* = "small," *florum* = "flower"; *pubescens* = "downy," covered with tiny hairs.

Common names: large yellow lady's slipper, yellow lady's slipper.

Description: Leaves 3–5, 5–20 cm (2–8") long, narrowly to broadly lance-shaped or nearly round with bases sheathing the stem, strongly veined, arranged alternately on the stem, green. **Stem** 20–70 cm (8–27½") or more tall, densely pubescent, green, supporting 1 or 2 flowers at the summit. **Flowers** large, up to about 15 cm (6") tall, 11½ cm (4½") wide, 6 cm (2⅖") long, with a yellow pouch-like lip, fragrance moderate to faint. **Sepals** and **petals** similarly colored, yellowish green with brown or reddish brown veining, dorsal sepal 3–8 cm (1⅕–3") long, lance-shaped, margin often wavy, erect or angled forward over the lip; lateral sepals united to form one blade similar in size and shape to the dorsal sepal but positioned behind the lip; petals 4–9 cm (1⅗–3½") long, long and slender and usually loosely spirally twisted, spreading and drooping; **lip** large, up to 5.4 cm (2 1/10") long, an elongated pouch open at the top, yellow on the outside, red to purple spotted inside.

Habitat: moist deciduous and coniferous forest, thickets, meadows, prairies, sometimes tundra and occasionally fens. Often in calcareous soils.

Flowering season: second week of May to second week of June (third week in May to first week in June).

Range: across most of the continent south of (and locally on) the tundra, from Newfoundland to Alaska. Absent from the central and southern prairies and the drier regions of the West, where it is spotty at higher elevations. It is found throughout the northeast.

Comments: Like the showy lady's slipper, the yellow lady's slipper has a reputation for bestowing a nasty dermatitis on overeager naturalists who come into contact with this species. The physical form of this variety of orchid is highly variable, depending on the environment. Sheviak (1995) has pointed out that plants growing in sunny, exposed sites in hilly calcareous soils may be small with flowers no larger than those of other varieties.

Courtesy Charles J. Sheviak, New York State Museum

Cypripedium parviflorum Salisbury var. **parviflorum**

Common name: small yellow lady's slipper.

Description: Leaves 3–5, broadly lance-shaped or nearly round with bases sheathing the stem, strongly veined, arranged alternately on the stem, green. **Stem** to 35 cm (13$^{4}/_{5}$″) tall, pubescent, green, supporting 1 or 2 flowers at the summit. **Flowers** about 5 cm (2″) tall, 9 cm (3½″) wide, 2.5 cm (1″) long, with a moderate to faint rose-like or pungent-musty fragrance. **Sepals** and **petals** similarly colored, uniformly crimson purple to dark reddish brown; resulting from parts being densely and minutely spotted; dorsal and united lateral sepals up to 4 cm (1$^{3}/_{5}$″) long; petals up to 5 cm (2″) long, yellow.

Habitat: rather dry to moist deciduous forest, often in markedly acidic soils.

Range: across the South to the Ozarks and the woodlands bordering the Missouri River, north to Missouri, northern Illinois, and across to southern New England. In the northeast var. *parviflorum* is found in the Southern Tier and mid-Hudson Valley of New York east to Massachusetts.

Flowering season: Sheviak conjectures that the blooming dates for this variety are similar to those of var. *pubescens*.

Comments: This is the rarest of the three varieties of yellow lady's slipper in the northeast. Many historical locations or populations no longer exist, but the search is on to determine its present status in the northeast.

Cypripedium parviflorum Salisbury var. **makasin** (Farwell) Sheviak

Etymology: *makasin* = "moccasin," as in native American footwear.

Common name: small yellow lady's slipper.

Description: Leaves 3–5, narrowly lance-shaped to nearly round with bases sheathing the stems, strongly veined, arranged alternately on the stem, green. **Stem** to 35 cm (13$^{4/5}$″) tall, somewhat pubescent, green, supporting 1 or 2 flowers at the summit. **Flowers** about 5 cm (2″) tall, 9 cm (3½″) wide, 2½ cm (1″) long, with an intensely sweet fragrance. **Sepals** and **petals** similarly colored, uniformly suffused with a dark reddish brown; dorsal and united lateral sepals up to 4 cm (1$^{3/5}$″) long; petals up to 5 cm (2″) long, spirally twisted; **lip** 1.5–2.9 cm ($^{6/10}$–1$^{1/10}$″) long, yellow.

Habitat: in the northeast, almost exclusively calcareous fen and *Thuja* swamp forests. Elsewhere also in moist calcareous deciduous or coniferous forest, thickets, and prairies.

Flowering season: third week in May to second week in June (fourth week in May to second week in June). Although a slightly later bloomer than var. *pubescens,* there is much overlap, and it is common to find both varieties in bloom on the same day.

Range: Most of the northeast and New York (except for northern Maine, Rhode Island, eastern Massachusetts and Connecticut, and Long Island), south to North Carolina, west through the Great Lakes region and the boreal forest. Becoming rare west of central Canada, and very rare in the mountains of the Pacific northwest.

Comments: Var. *makasin* is the more common of the two small-flowered yellow lady's slippers in the area covered by this book.

Cypripedium reginae Walter

Etymology: *reginae* = "of the Queen," in reference to the royal beauty of the bloom, hence the common name of lady's slipper of the Queen. We can only speculate as to which other species might represent the king.

Common names: lady's slipper of the Queen, Queen's lady's slipper, showy lady's slipper.

Description: Leaves 3–7, 7.5–20 cm (3–8″) long, broadly egg-shaped with bases sheathing the stem, strongly veined, pubescent, arranged alternately on the stem, green. **Stem** usually 38–85 cm (15–33½″) tall, stout, densely pubescent, green, supporting 1 to 2 or rarely 3 flowers at the summit. **Flowers** large and showy, about 8 cm (3″) high, 8–9 cm (3–3½″) wide, 4–5 cm (1⅗–2″) long, with a rose pink pouch-like lip. Dorsal **sepal** 3–4.5 cm (1⅕–1¾″) long and almost as wide, nearly round, angled upward at 45–55° above the lip, white; lateral sepals united to form one nearly round blade positioned behind the lip, slightly smaller than the dorsal sepal, white; **petals** 2.5–4 cm (1–1⅗″) long, oblong, spreading, white; **lip** 2.5–5 cm (1–2″) long, a nearly globular pouch open at the top, magenta to rose pink with thin vertical white stripes on the outside, purple-pink spotted inside.

Habitat: edges of fens, swamps, moist meadows, and woods.

Flowering season: the second week of June to third week of July (the second week of June to third week of July).

Range: from Newfoundland south to New Jersey, west to Missouri and southeastern Alberta. It is found throughout the northeast.

Comments: The lady's slipper of the Queen is regarded as the largest and showiest orchid flower in northeastern North America. It is still surprisingly abundant in many parts of its range, especially considering man's propensity to plunder prominent orchids to near extinction. I cannot speak for all parts of its range, but near my home in the mountainous region of New York it owes at least part of its protection to legions of tiny blackfly warriors. Blackfly populations and the Queen's blooms peak simultaneously, and at that time the aggressive little flies with the big bites drive all but the most dedicated naturalists from the fens and moist meadows! Unlike the orchid flowers, the general form of this species advises avoidance, for physically the stem and leaves closely mimic those of the poisonous false hellebore, *Veratrum viride,* which often grows nearby. In the case of the Queen this visual threat is not to be too quickly dismissed, for our tallest orchid has a reputation for punishing human handlers with a nasty dermatitis. Baldwin (1884) disagreed with this, believing that the victims had inadvertently touched some nearby poison ivy.

Specimens with ivory white flowers are not uncommon. Although today these white blooms are classified as forma *albolabium,* in early American history some naturalists believed they represented a separate species, which they named *Cypripedium album.*

Cypripedium reginae forma *albolabium*

Calypso

Calypso bulbosa (Linnaeus) Oakes

Etymology: *Calypso* = from the name of a beautiful nymph in the Odyssey; *bulbosa* = "bulb-like," referring to the underground portion of this plant.

Common names: calypso, fairy slipper.

Description: Leaf solitary, blade 2.5–6 cm (1–2⅖″) long and nearly as wide, stalk 2.5–6 cm (1–2⅖″) long; blade roundish to egg-shaped, strongly pleated with parallel veins, dark green to bluish-green. **Stem** 7.5–15 or up to 20 cm (3–6 or up to nearly 8″) tall, wrapped in 2–3 leaf-like sheaths, purplish. **Flower** solitary, about 3.5–5 cm (1⅖–2″) tall and 3–4 cm (1⅕–1⅗″) wide. **Sepals** and **petals** similar, 15–25 mm (⅗–1″) long, narrowly lance-shaped, sometimes slightly twisted, all spreading upward or outward above the lip, pale purplish to rose; **lip** about 15–25 mm (⅗–1″) long, saccate, pendent, lower portion of lip broad, usually white or sometimes tinged with pink, with a few purplish spots on the margins of the opening and a conspicuous tuft of golden yellow hairs just below, inside of lip vertically streaked with white and purplish markings.

Habitat: grows in rich humus in moist, mossy woodlands. In the northeast it is usually associated with swamps of white cedar *(Thuja occidentalis).*

Flowering season: third week of May to second week of June (extirpated in New York, historically fourth week of May to first week of June).

Range: Primarily a Canadian-Alaskan orchid, it has been recorded as far south as the Adirondack Mountains in the east and New Mexico and Arizona in the west. In the northeast *Calypso* has dramatically decreased in both range and numbers in this century, but within recent times it has been recorded in most of Maine, northern Vermont and New Hampshire, and northeastern New York.

Comments: The calypso genus has only one member, but various varieties of that single species are found throughout the northern hemisphere in North America, Europe, and Asia. Despite its extensive range, calypso does not survive close contact with man. In the northeast today this once well-established orchid has been reduced to a handful of remote colonies.

The common name of calypso is taken from a nymph in the Odyssey. Like the orchid, she was known for her great beauty and her habit of living far from the haunts of man. The alternate common name of fairy slipper is a fanciful interpretation of the colorful drooping and inflated lip, with its golden yellow "powder-puff" decorations across the upper surface. Writing in 1929, Morris and Eames played on this image by comparing the lip to a "wide, pointed shoe hung up by the heel with the opening in front." In 1975 Luer embellished the beauty of this image by imagining the spreading sepals and petals as decorative rose-colored ribbons.

Although it was later believed that calypso had no close relatives, in the 1700s the early botanist Linnaeus was of the opinion it was one of the lady's slippers, basing his belief on the greatly inflated lips he observed in both groups. Today *Calypso* has been included in the winter-leaf orchids. The three members of this group, calypso, putty-root, and the cranefly orchid, all produce a single leaf in September (calypso, September 2; putty-root, September 9; and cranefly, September 14; according to Baldwin 1884) that survives the rigors of winter only to shrivel and die the following spring. Calypso is the earliest blooming of these three orchids, and it is the only member of this group that typically still has the leaf present in good condition at flowering time. (Putty root leaves may be present at flowering, but in a declining yellowing condition.) Calypso is a delicate orchid that requires cool, moist habitats to ensure its survival. Scientists are now tracking calypso and a number of other plants with similar temperature sensitivities in an effort to gauge the effect of greenhouse gases in our atmosphere.

Although the calypso typically displays only a single bloom, specimens sporting two flowers are occasionally observed.

Group 2

Calopogon
Arethusa
Pogonia

Calopogon

Calopogon tuberosus (Linnaeus) Britton, Sterns, and Poggenberg

Etymology: *Calopogon: calo* = "beautiful," *pogon* = "beard," beautiful beard referring to the yellow-tipped hairs on the lip; *tuberosus* = "tubered," referring to the enlarged underground stem.

Common names: grass-pink, calopogon.

Description: Leaf most often solitary but occasionally several, 15–30 cm or up to 50 cm (6–12 or up to 20″) long, long and slender, basal, green. **Stem** 30–46 or up to 120 cm (12–18 or up to 47″) tall, smooth, green. **Inflorescence** about ⅕–¼ of the total stem length, usually 10–38 cm (4–15″) long, with 3 to 15 or up to 25 loosely arranged flowers. **Flowers** about 30–35 mm (1⅕–1⅖″) tall and wide, all parts pink to rose-pink or magenta. Central **sepal** 21–25 mm (⅘–1″) long, narrowly lance-shaped; lateral sepals about 15–20 mm (⅗–⅘″) long, somewhat egg-shaped, dorsal sepal curved downward and forward in the position occupied by the lip in most orchids, lateral sepals spreading outward, the tips of all sepals curved slightly forward; **petals** about 18–23 mm (7/10–9/10″) long, oblong, slightly constricted near the center, spreading with tips curving forward; **lip** about 14–18 mm (1/2–7/10″) long, long and narrow but flaring and becoming triangular near the tip, pink to magenta with a conspicuous tuft of white to yellow-tipped hairs near the tip, growing erect in the position occupied by the dorsal sepal in most orchids.

Habitat: open fens, moist meadows, floating mats along ponds and rivers, occasionally growing quite close to ocean beach areas.

Flowering season: third week of June to first week of August (fourth week of June to third week of July, a few into the fourth week of July).

Range: from Newfoundland south to Florida, west to southeastern Manitoba and eastern Texas. It is also found in Cuba and the Bahamas.

Comments: Calopogon is something of a trickster in the wildflower world, for compared with other orchids, its flowers grow upside down! The central sepal occupies the lowest position on the bloom while the lip is placed above the other floral parts, just the opposite of what we consider to be a typical orchid arrangement. Calopogon's trickery also extends to its reproductive rituals. To the local bees and other insects, the tuft of yellow-tipped hairs on the erect lip must appear similar to those found in nearby arethusa and pogonia, both suppliers of nectarous meals. A bee landing on calopogon may attempt to scale the lip in search of the promised sweet liquid, but finds none. Instead, the bee discovers that not only can the narrow lip not support its weight but it collapses forward, dumping the duped insect onto its back. When the bee lands in that unflattering position, a cluster of sticky pollen grains are attached to its back. The bee then transports this sticky pollen to other calopogons as it continues its endless quest for floral nectar.

Occasionally one hears of the calopogon lip having the power to snap down on the visiting bee in the same manner as a mouse-trap shutting or a Venus flytrap closing. Such claims may be as colorful as the lip itself, but the reality appears to be that it is the weight of the bee that is alone responsible for the downward collapse of the lip.

Calopogon frequently lives in the company of two similarly colored bog orchids, arethusa and the rose pogonia. This trio comprise three of the all-time favorite orchids, for all are beautiful and easy to identify. If we travel south to Florida we will encounter three or four smaller species of *Calopogon,* which together with *C. tuberosus* comprise the entire genus. Although not common, pure white calopogon blooms are occasionally observed.

Arethusa

Arethusa bulbosa Linnaeus

Etymology: *Arethusa:* from the name of a beautiful nymph in Greek mythology; *bulbosa* = "bulb-like," referring to the underground portion of this plant.

Common names: arethusa, dragon's mouth, moss nymph, wild pink.

Description: Leaf solitary, 10–15 cm or up to 20 cm (4–6″ or up to 8″) long, long and slender, appearing as a short (about 4 cm or 1⅗″) leaf or bract and only elongating after the flowering season, positioned ⅓–½ way up the stem, green. **Stem** 13–30 or up to 40 cm (5–12 or up to 16″) tall, wrapped in 2 to 3 leaf-like sheaths, color variable from yellow-green to dark green to purplish. **Flower** solitary, about 38–50 mm (1½–2″) tall and about 30 mm (1⅕″) wide. **Sepals** 3–4 cm (1⅕–1⅗″) long, narrowly lance-shaped, dorsal sepal erect, lateral sepals erect and flaring slightly backward, rose magenta; **petals** about 3 cm (1⅕″) long, rose magenta; dorsal sepal and petals erect but curving slightly forward to form a loose hood over the lip; **lip** about 3 cm (1⅕″) long, oblong and indistinctly 3-lobed, erect at the base then folded forward and pendent for half its length, basal portion of lip broad with an irregular wrinkled margin, white to pinkish-white with pinkish margins and purplish markings throughout, and a tuft of yellowish to white hairs along the center.

Habitat: in the northeast arethusa is usually found growing from moss in moist and sunny sphagnum fens.

Flowering season: third week of May to second week of July (second to fourth week of June).

Range: from Newfoundland south to New Jersey, westward around the Great Lakes. There are also isolated populations in the mountains of western North Carolina, presumedly survivors of ancient glacial times. Historically arethusa was found in fens throughout the northeast, but within the last century it has vanished from many localities.

Comments: Arethusa appears to be the sole American representative of a very limited orchid group. Its only close relation is found in Japan. This beautiful orchid is one example of a plant whose genus name and common name are one and the same: *Arethusa*. This species is one of the very few orchids whose original species name, which dates back to 1753, has remained intact. Arethusa in ancient mythology was a beautiful nymph turned into a flowing spring to protect her from an overly amorous pursuer, hence the older alternate common name of moss nymph. I especially like the common name of dragon's mouth, for arethusa is clearly one of those orchids into whose physical form we enjoy projecting our own mental associations. In arethusa we see a fiery-red water dragon raising from some prehistoric fen. The spreading sepals represent some dinosaur-like bony plates, the petals curve forward to form the head, and the lip hangs down as a toothy jaw open to spew an imaginary fireball at any mortals who dare come too close!

Arethusa is sometimes compared to calypso, for both bloom early, share a crown of beautiful erect sepals, and have a "hairy" pendent lip. The two do not, however, share the same habitat, for calypso prefers shaded cedar swamps, while arethusa is a sun-loving resident of open sphagnum fens. They also differ in other elements of natural history and life cycle, for in arethusa the single long and slender leaf is usually not fully developed until after the flowers fade. Two unusual color forms of arethusa have also been noted. In one the blooms are pure white, in the other lilac-lavender. Although far from common, single stalks bearing two blooms are occasionally observed.

Pogonia

Pogonia ophioglossoides (Linnaeus) Jussieu

Etymology: *Pogonia* = "beard," referring to the bearded lip; *ophioglossoides*: ophis = "snake," glossa = "tongue," eidos = "like," "resembling an *Ophioglossum*," the adder's tongue fern, whose leaf superficially resembles that of this species.

Common names: rose pogonia, snake mouth, adder's mouth.

Description: Leaf usually solitary, 1–7.5 or up to 10 cm ($^2/_5$–3 or up to 4″) long, oblong to egg-shaped, usually growing from a point less than halfway up the stem (occasionally with another long-stalked leaf growing from the base of the stem), green. **Stem** 20–35 cm (8 to nearly 14″) tall, green, becoming darker towards the base, bearing typically one but occasionally two or even three flowers. **Flower** about 32–38 mm (1¼–1½″) tall, wide and long, pink, fragrant. **Sepals** similar, about 21–23 mm ($^4/_5$–$^9/_{10}$″) long, narrowly lance-shaped, dorsal sepal erect or curved forward, lateral sepals spreading but curved forward, pink; **petals** about 21–23 mm (⁴/₅–⁹/₁₀″) long, paddle-shaped, curved forward over the lip, pink; **lip** about 21–23 mm (⁴/₅–⁹/₁₀″) long, narrowly oblong with a heavily fringed margin and a tuft of yellowish or greenish thick hair-like projections in the center, pink with reddish pink margins and veining.

Habitat: open fens, river banks, moist meadows, and roadside ditches.

Flowering season: third week of June to third week of July, a few into August (fourth week of June to second week of July, a few third week of July).

Range: from Newfoundland to Florida, west to the Mississippi River Valley. It is found throughout the northeast.

Comments: In the latter part of the last century, the genus *Pogonia* was viewed as a more diverse and inclusive group than it is today. All the members of this genus had lips ornamented with tufts or ridges of often brightly colored hair-like structures. *Pogonia* means "beard" and refers to this feature. At that time four northeastern orchids were included in this genus: the rose pogonia, three-birds (also known as the nodding pogonia), and the large and small whorled pogonias.

As botanists better understood the physical intricacies of orchids, the genus *Pogonia* was split into several separate genera, and the four species listed above now account for three separate genera. The rose pogonia retained the genus name *Pogonia,* three-birds joined with some southern relatives to become *Triphora,* while the two whorled pogonias became viewed as the only members of an exclusively eastern North American genus, *Isotria*. Another species once listed under *Pogonia* but now a member of a separate genus is *Cleistes divaricata,* the large and beautiful spreading pogonia, which is found from New Jersey south to Florida.

The rose pogonia is a favorite among orchids, with much to recommend it. It is beautiful in a way only orchids can be, it is quite common, and it is found in sunny open places. As has already been pointed out, the genus *Pogonia* was a somewhat more diverse taxonomic entity at the turn of the century. Today the rose pogonia is the sole American representative that retains the original name of this now very limited genus, with the only other members of the genus in Japan and China.

When Thoreau wrote "Summer" in 1884 he had some harsh words regarding what he considered the offensive odor of this otherwise beautiful species. His opinion stands in stark contrast to that of others who find the fragrance delightful, reminiscent of red raspberries or sweet violets. Writing in 1905, Gibson suggested a possible resolution of these divergent views, observing that this flower emits a favorable fragrance when fresh, but one that becomes disagreeable as the bloom withers.

The rose pogonia typically sports a single flower on each stalk, but two or even three blooms are not unknown, especially in the southern part of its range. The typical color of the rose pogonia is a pleasing pink, but pale pink to white specimens have been regularly observed. The pogonia's flowering season usually comes after arethusa and overlaps with calopogon, but in years with a late spring, such as occurred in 1994, all three of these fen loving orchids may be found in bloom within sight of each other! More than any other orchid, the rose pogonia has highly variable flower sizes, with some fens occasionally producing flowers nearly twice the typical dimensions.

Group 3
Isotria

Isotria
The Whorled Pogonias

Etymology: *Isotria* = "with three equal parts," probably referring to the similarly sized and shaped sepals.

Isotria can be quickly recognized by their whorl of more or less egg-shaped leaves, a distinct foliage pattern not shared by any other northeastern orchids. Immature or nonflowering plants are sometimes confused with *Medeola virginiana*, the Indian cucumber-root. An examination of the stalks can be helpful in this situation, for those of *Medeola* are thin and minutely hairy, those of *Isotria* are stouter, smooth, and hollow.

Key to Isotria
(continued from page 8)

I. Sepals very long and slender, much longer than the petals, brownish to purplish.
I. verticillata → page 49.

II. Sepals and petals of nearly the same length, yellowish green.
I. medeoloides → page 51.

Isotria verticillata (Muhlenburg ex Willdenow) Rafinesque

Etymology: *verticillata* = "whorled," referring to the leaves.

Common names: whorled pogonia, large whorled pogonia.

Description: Leaves 5–6, 2.5–9 cm (1–3½") long but often smaller at flowering time, more or less egg-shaped to lance-shaped, arranged in a whorl at the top of the stem, green. **Plant** 13–30 cm (5–nearly 12") tall. **Stem** smooth, purplish, ending at the whorl of leaves. **Flower** usually solitary, rarely double, about 65–75 mm (2½–3") tall, 25 mm (1") wide, 19 mm (¾") long, growing from a long 25–38 mm (1–1½") greenish stem-like ovary that visually appears to be an extension of the stem above the leaves. **Sepals** about 40–50 mm (1³/₅–2") long, very long and slender, spreading, brownish to purplish; **petals** 20–25 mm (⁴/₅–1") long, lance-shaped, growing forward over the lip, greenish yellow to greenish; **lip** about 20 mm (⁴/₅") long, somewhat wedge-shaped, 3-lobed, lateral lobes growing erect and streaked with purple, central lobe off-white to greenish yellow with a wavy margin.

Habitat: dry to damp hardwoods to mixed forests, also along the mossy edges of fens.

Flowering season: third week of May to second week of June (fourth week of May to first week of June).

Range: from southern Maine west to Michigan, south through most of Georgia and Louisiana. Found throughout the northeast except for northern New York, Vermont and New Hampshire, and most of Maine.

Comments: The long, slender, and spreading sepals give the whorled pogonia a decidedly tropical appearance. This small and delicate orchid springs from long (sometimes over 12″) roots. Being unaware of the invisible yet extensive root system, misguided naturalists who attempt to transplant this species invariably damage the roots and succeed only in sealing the flower's doom. (Several additional factors such as a narrow range of acceptable environmental conditions also preclude the possibility of successfully transplanting this and most other orchids.) These roots grow horizontally under the surface and often sprout new stems from their tips. It is fortunate that this beautiful orchid has evolved this trick of vegetative propogation, for each plant blooms only once every few years, and only a few flowers succeed in producing viable seed. Because several plants may grow from each set of roots, the whorled pogonia is most often found in small colonies of plants representing all sizes and stages of growth.

Isotria medeoloides (Pursh) Rafinesque

Etymology: *medeloides* = "like a *Medeola*"; *Medeola virginiana*, the Indian cucumber-root, has a similar whorl of leaves.

Common name: small whorled pogonia.

Description: Leaves 5–6, 2.5–8 cm (1–3″) long, more or less egg-shaped, arranged in a whorl at the top of the stalk, green. **Plant** 7.5–25 cm (3–nearly 10″) or more tall. **Stem** smooth, whitish green, ending at the whorl of leaves. **Flowers** solitary or double, about 1.5–3.5 cm (³/₅–1²/₅″) tall, 1.5–2 cm (³/₅–⁴/₅″) wide and long, growing from a short 1.5–2 cm (³/₅–⁴/₅″) pale green stem-like ovary that visually appears to be an extension of the stem above the leaves. **Sepals and petals** similar, about 20 mm (⁴/₅″) long with petals slightly shorter, narrowly lance-shaped, sepals spreading to form a triangular frame around the petals, petals growing forward over the lip, pale green to yellowish green; **lip** about 15 mm (³/₅″) long, somewhat wedge-shaped with a wavy margin, 3-lobed, lateral lobes growing erect and greenish tinged, central lobe whitish with a number of yellowish green knobby growths at the base.

Habitat: in rich leaf litter in hardwood forests.

Flowering season: fourth week of May to second week of June (first to second week of June).

Range: very rare and local. States known to have had isolated populations include Illinois, Missouri, Michigan, North Carolina, Virginia, Maryland, Pennsylvania, New Jersey, and all of the states in the northeast, but some of these stations no longer exist.

Comments: By all accounts one of the rarest orchids in the eastern United States. Because of its small size and inconspicuous coloration, it is likely that additional yet-to-be-discovered colonies exist. Difficulty in locating the small whorled pogonia is compounded by its reported ability to exist underground for many years without putting up a flowering stem. In New York, for example, the last known population has not produced above-ground stems in many years, leading to speculation as to whether or not the species still exists in that state.

Group 4

Goodyera
Spiranthes

Goodyera
The Rattlesnake Plantains

Etymology: *Goodyera:* in honor of John Goodyer, an English botanist who lived from 1592–1664.

Only four of the twenty-five or so *Goodyera* species that are found around the world inhabit the western hemisphere. In the northeast, we are fortunate to share this area with all four of these species, although one, *G. oblongifolia,* is found only at the very northern tip of the region.

As is often the case, the names given to these plants by those who came before give us an indication of their physical character. The name rattlesnake does not imply any venomous or toxic trait; it is instead a reference to the distinctive reticulations (net-like patterns) on the leaves that are the hallmark of this genus. Those same imaginative minds that saw snake scales in the leaf pattern also established and perpetuated the herbal myth that this orchid could be employed to counteract the effects of snakebite. The reality, of course, is that any poor soul who attempts this cure will be committing the double blunder of (1) wasting valuable time on a dubious medical practice, and (2) the murder of an innocent orchid!

The name plantain also refers to the leaves, whose general outline is reminiscent of the common weed plantains, *Plantago* spp., which despite our best efforts to remove them continue to inhabit and often infest our lawns.

It is unusual among orchids to find species where the distinctive beauty of the leaves rivals and even surpasses that of the flowers, but this is just the case with the rattlesnake plantains. After viewing these orchids, it is easy to understand why other common names refer only to the leaves, for their unique colorations and patterns quickly distract us from the attractive yet understated and undersized white blossoms. Beauty, unfortunately, is often the downfall of wild plants because it attracts the unwanted attentions of man. Large numbers of rattlesnake plantains have been killed by being transplanted into tiny terrariums, where their uniquely marked evergreen leaves are highly prized. This practice has long been carried out by both commercial exploiters and individual collectors, but with increased awareness it is hoped that this misguided effort will soon die out.

In the wild, rosettes of evergreen leaves (unusual in northern orchids) grow from the rhizomes, underground stems that grow horizontally just under or upon the soil. Since each rhizome may sport several rosettes of leaves, it is not unusual to find rattlesnake plantains growing in groups or even large colonies. *Goodyera pubescens* is perhaps the species most often found in large numbers. The leaves of the rattlesnake plantains are distinctive enough that an educated guess, if not a positive identification, may be made of plants lacking flowers. For that reason, this guide also includes a key to identifying *Goodyera* by leaf characteristics.

As beautiful as the leaves are, it would be our loss if we allowed them to distract us from ever getting around to closely examining the flowers. Viewed under a hand lens, they look like tiny perfect porcelain milk pitchers, with the lip/spouts ready to pour out a tiny glass of fragrant refreshment to any woodland spirits stopping to set a picnic on the checkered blanket below.

One important fact to consider is that when dealing with the rattlesnake plantains, it is desirable and usually more accurate to base identifications on several characteristics. One of the commonest rattlesnake plantains in the northeast, *G. tessellata,* is believed to be a hybrid resulting from an ancient cross between *G. repens* and *G. oblongifolia,* two quite distinct *Goodyera* species. As hybrids frequently exhibit a confusing array of traits that may overlap with those of either parent, it is often necessary to consider several characteristics to confirm an identification, especially of this species.

Even understanding and accepting these difficulties in advance, it is hoped that the following key will provide some assistance in identifying *Goodyera*. Please keep in mind that all the rattlesnake plantains are to some degree variable, and even though knowledge and experience leads me to believe that these are the typical characteristics of these orchids in the northeast, I caution the reader that most if not all of these entries should be preceded by the cautionary word *usually!*

Key to Goodyera, Based on Floral Characteristics
(continued from page 8)

I. Flowers nearly spherical when viewed from the side.
   A. Tip of lip triangular, deeply to shallowly downcurved.
      1. • Flowers about 6 mm (¼") long.
         • Lip somewhat saccate to cup-shaped, tip somewhat to shallowly downcurved.
         • Flowers loosely spiraled to one-sided on the stem.
           G. tesselata → page 58.
      2. • Flowers about 3 mm (⅛") long.
         • Lip narrowly saccate, tip deeply downcurved.
         • Flowers one-sided on the stem.
           G. repens → page 60.
   B. Tip of lip bluntly triangular, deeply downcurved.
      3. • Flowers about 5 mm (⅕") long.
         • Lip broadly saccate to scrotum-like, tip deeply down curved.
         • Flowers forming a densely cylindrical cluster.
           G. pubescens → page 62.

II. Flowers somewhat pear-shaped when viewed from the side.
   A. Tip broad and tongue-shaped, slightly downcurved.
      4. • Flowers about 9 mm (⅓") long.
         • Lip shallowly cup-shaped.
         • Flowers one-sided to loosely spiraled on the stem.
           G. oblongifolia → page 64.

Key to Goodyera, Based on Leaf Characteristics.
(continued from page 8)

I. Leaves typically lacking a whitish stripe down the center.
   A. • Reticulations wide, occasionally faint.
      • Leaves 2–8 cm (⁴/₅–3″) long, usually pale green or pale bluish green.
         1. *G. tesselata* → page 58.
   B. • Reticulations wide, occasionally faint or absent.
      • Leaves 1–3 up to 4 cm (²/₅–1¹/₅ up to 1³/₅″) long, usually dark green or sometimes lighter.
         2. *G. repens* → page 60.
II. Leaves with a whitish stripe down the center.
   A. • Reticulations fine, completely interconnected.
      • Leaf 2.5–5 or up to 9 cm (1–2 or up to 3½″) long, bluish green.
         3. *G. pubescens* → page 62.
   B. • Reticulations sometimes absent, cross-bars often not interconnected.
      • Leaf 3–7 or up to 11 cm (1¹/₅–2³/₄ or up to 4³/₁₀″) long, dark green to pale bluish green.
         4. *G. oblongifolia* → page 64.

Goodyera tesselata Loddiges

Etymology: *tesselata* = "checkered," referring to the reticulations on the leaves.

Common names: tesselated rattlesnake plantain, Loddiges' rattlesnake plantain.

Description: Leaves 3–8, 2–8 cm (⁴/₅–3″) long, broadly lance-shaped, forming a basal rosette, pale green to pale bluish green with wide whitish reticulation, evergreen. **Stem** 13–25 or up to 35 cm (5–10 or up to nearly 14″) tall, pubescent, usually with a few tiny bracts. **Inflorescence** about ⅓ of the stem, usually 5–13 cm (2–5″) long, with 15–30 or up to 40 or more flowers, flowers loosely spiraled to one-sided on the stem. **Flowers** about 6 mm (¼″) tall, 6–8 mm (¼–³/₁₀″) wide, and 6 mm (¼″) long, nearly spherical when viewed from the side, white. **Sepals** and **petals** similar, about 6 mm (¼″) long; the petals slightly shorter, lance-shaped, the petals slightly narrower; dorsal sepal and petals converging to form a sheltering hood over the lip, lateral sepals curving forward along the lip but often spreading outward at about a 45° angle; lip about 4 mm (1/6") long, somewhat saccate to cup-shaped, tip triangular, somewhat to shallowly downcurved.

Flowering season: third week of July to first week of September (first to third week of August).

Habitat: dry to damp areas in coniferous, hardwood, or mixed forests.

Range: Tesselated rattlesnake plantain is a northeastern species that is found around all but the southernmost regions of the Great Lakes, and east to Newfoundland and Pennsylvania. It is found throughout the area covered by this work, excluding only Cape Cod and southern Connecticut.

Comments: The tesselated rattlesnake plantain is often first identified by the pale bluish green color of its leaves, although occasionally a specimen is encountered with dark green leaves. *Goodyera tesselata* is believed to have its origins as an ancient hybrid between *G. repens* and *G. oblongifolia,* and in most vegetative and floral characteristics it is intermediate between the two original parents. Jacquelyn A. Kallunki carried out a thorough exploration of these traits, which formed the basis of her excellent article (1976) on the subject. *G. tesselata* was the last of the *Goodyera* species to be widely recognized and accepted. For this reason, descriptions of the tesselated rattlesnake plantain are lacking in many early orchid books.

Goodyera repens (Linnaeus) R. Brown

Etymology: *repens* = "creeping," referring to the rhizomes.

Common name: Lesser rattlesnake plantain.

Description: Leaves 3–7, 1–3 up to 4 cm (²/₅–1¹/₅ up to 1³/₅″) long, egg-shaped, forming a basal rosette, dark green, usually with wide whitish reticulations which may be absent in some specimens, evergreen. **Stem** 13–25 cm (5–10 inches) tall, pubescent, usually with a few tiny bracts. **Inflorescence** ¹/₄–¹/₃ of the stem, usually 3–5 cm (1¹/₅–2″) long, with 10–15 or up to 20 or more flowers, flowers usually arranged on only one side of the stem. **Flowers** about 4 mm (¹/₆″) tall, 3–4 mm (¹/₈–¹/₆″) wide, and 3 mm (¹/₈″) long, nearly spherical when viewed from the side, all parts white, sometimes with green or pink tints. **Sepals** about 3–4 mm (¹/₈–¹/₆″) long, somewhat egg-shaped; **petals** 3–4 mm (¹/₈–¹/₆″) long, narrowly egg-shaped, dorsal sepal and petals converging to form a sheltering hood over the lip, lateral sepals curving forwards along the lip but with spreading tips; **lip** about 3 mm (⅛″) long, narrowly saccate, tip triangular and deeply downcurved.

Flowering season: second week of July to third week of August (first to second week of August).

Habitat: usually in moist coniferous forests, frequently growing on moss.

Range: Lesser rattlesnake plantain can be found living on all the northern continents. In the eastern hemisphere it inhabits a wide band stretching from Japan and eastern Russia across Asia and through most of Europe to northern Britain. In the western hemisphere it is found from Newfoundland to Alaska, and south to the mountains of Tennessee in the east and New Mexico in the west. It is found throughout the northeast.

Comments: Many people have commented that the lesser rattlesnake plantain looks very much like a small downy rattlesnake plantain, and that the two can be easily confused. The lesser rattlesnake plantain can be identified by its smaller size, one-sided flowering spike, wider leaf reticulations, and a proportionately longer tip protruding from the lower lip. The lesser rattlesnake plantain is one of the species featured by Darwin in his two books on Orchid pollination (1884 and 1877).

Goodyera pubescens (Willdenow) R. Brown

Etymology: *pubescens* = "downy," referring to the covering of small soft hairs on the stalk.

Common name: downy rattlesnake plantain.

Description: Leaves 3–8, 2.5–5 or up to 9 cm (1–2 or up to 3½″) long, oblong to broadly lance-shaped, forming a basal rosette, bluish green, usually with a conspicuous white stripe along the central vein, the blade area marked with a fine white reticulation, evergreen. **Stem** 15–50 cm (6–20″) tall, stout, pubescent, usually with a few tiny bracts. **Inflorescence** ¼–⅓ of the stem, usually 3–12 cm (1¼–4½″) long, cylindrical, with 20–60 or up to 80 densely arranged flowers. **Flowers** about 5 mm (⅕″) high, 6 mm (¼″) wide, and 5 mm (⅕″) long, nearly spherical when viewed from the side, all parts white, sepals often with greenish tips and tints. **Sepals** similar, about 5 mm (⅕″) long, egg-shaped; **petals** about 6 mm (¼″) long, paddle-shaped; dorsal sepal and petals converging to form a sheltering hood over the lip, lateral sepals curving forward along the lip but with spreading tips; **lip** about 4 mm (⅙″) long, broadly saccate to scrotum-like; tip short, bluntly triangular and deeply downcurved.

Habitat: dry to damp areas in hardwood, coniferous, or mixed forests.

Flowering season: third week of July to first week of September (first to third week of August).

Range: The downy rattlesnake plantain is found from central Maine south to northern Georgia, west to the Mississippi River and the western edge of the Great Lakes. It is found throughout the northeast except for northern Maine and Cape Cod.

Comments: Do not be misled by the Latin name of *pubescens* (covered with tiny hairs) used to describe the stem of the species. To some degree the stems of all four rattlesnake plantains sport a covering of fine hairs.

Goodyera oblongifolia Rafinesque

Etymology: *oblongi* = "oblong," *folia* = "leaf."

Common name: Menzies' rattlesnake plantain.

Description: Leaves 3–7, 3–7 or up to 11 cm (1⅕–2¾ or up to 4³⁄₁₀″) long, oblong to egg-shaped, forming a basal rosette, dark green to pale bluish green, usually with a conspicuous white stripe along the central vein, the blade area may or may not be marked with a white reticulation, evergreen. **Stem** 20–38 or up to 45 cm tall (8–15 up to nearly 18″), pubescent, usually with a few tiny bracts. **Inflorescence** ¼–⅓ of the stem, usually 8–13 cm (3–5″) long, with 12–24 or up to 30 or more flowers, flowers usually densely one-sided or occasionally loosely spiraled on the stem. **Flowers** about 8 mm (³⁄₁₀″) tall, 7–8 (¼–³⁄₁₀″) wide, and 9 mm (⅓″) long, somewhat pear-shaped when viewed from the side, white, often with some greenish coloration on the upper parts. Dorsal **sepal** 8–10 mm (³⁄₁₀–⅖″) long, long triangular; lateral sepals 7–8 mm (¼–

³⁄₁₀″) long, somewhat narrowly egg-shaped; **petals** 8–10 mm (³⁄₁₀–²⁄₅″) long, somewhat tear-shaped; dorsal sepal and petals converging to form a sheltering hood over the lip, lateral sepals with recurved tips; **lip** about 5 mm (⅕″) long, shallowly cup-shaped; tip broad and tongue-shaped, only slightly downcurved.

Habitat: dry to damp areas in conifers, hardwoods, or mixed forests.

Flowering season: first to fourth week of August.

Range: Menzies' rattlesnake plantain is found from the Atlantic to the Pacific in most of the southern Canadian provinces, and up into southern Alaska. It is found along much of the American west coast, and in mountainous terrain in the western states south into Mexico. In the northeast its range is very limited, being found only in the northernmost portions of Maine.

Comments: Menzies' rattlesnake plantain can best be identified by the distinctive white stripe along the center of each leaf, by the large pear-shaped flowers (the flowers of the other three rattlesnake plantains are more spherical in shape), and the blunt tongue-shaped tip of the lip. At one time this species was classified as *Goodyera menziesii,* in reference to an early describer of this plant. Another early classification of this species was *Goodyera descipiens, decipiens* referring to the deceptive physical characteristics of this species, which in its unreticulated form had been mistaken for a *Spiranthes.*

Spiranthes
The Ladies' Tresses

Etymology: *Spiranthes* = "spiraled blossom," an extremely accurate description of the flowering habits of this genus. The now obsolete genus name of *Gyrostachys* (round or turning spike) also referred to the manner in which the flowers spiral around the stem.

Spiranthes are some of the most beautiful of our "miniature" wild orchids. It is always a singular pleasure to come across unexpectedly a colony of the gracefully spiraled little white blossoms that are the hallmark of this genus. The origin of the common name of ladies' tresses has been lost as some old and unrecorded bit of botanical history, leaving us free to speculate on the meaning. The most romantic and widespread opinion suggests the ancient art of braiding flowers into the hair of maidens, and there is no doubt that the flowering stem of many ladies' tresses does resemble a braid-like pattern.

Spiranthes was once thought to be a giant genus with over 300 species, but today most of these have been reassigned to other groups, leaving *Spiranthes* as a much smaller genus with approximately 30 members. About two dozen species are thought to inhabit North America, and of these eight are found in the northeast. Within this area, *Spiranthes* are recognized by having numerous tubular white to creamy white flowers spiraled about a slender stem. The flowers tend to be quite small, with even the largest approaching only one-half inch in length. The leaves may be long and slender or broad and egg-shaped. The presence or absence of leaves during flowering is an important diagnostic trait in some species.

If there is one characteristic for which this genus is known, it is the confusing similarity that exists among the species. Morris and Eames (1929) light-heartedly referred to the "embarrassing wealth" of having too many species that are difficult to differentiate. So many species whose differences appear so inconspicuous to the human eye have led to a number of perplexing identification issues, some which have yet to be resolved. A white-throated form of *S. praecox* has been reported from Long Island several times. (Typically *S. praecox* has a white lip with several linear green markings.) Although these reports have not been confirmed, they can be considered at least possible, for *S. praecox* is found in nearby New Jersey. Far less likely, *S. praecox* has been reported as existing as far inland as Oneida County in New York. Whether or not the various species of this genus hybridize is another well-discussed issue. Although some plants previously thought to be hybrids have turned out not to be, examples of a *S. gracilis/S. vernalis* hybrid, *S.* × *intermedia* Ames, have been reported from Long Island. Most orchidologists are just as happy that the various members of this genus do not freely cross with each other, for identifications of *Spiranthes* can be difficult enough without having to contend with an additional "hybridization headache."

Like all orchid books, this work will include a key to assist the reader in species identification. Like other orchid books, however, we must caution that any key to this genus automatically will be flawed by the nature of the orchids observed. Every characteristic used here is variable to some degree, and those attributes that are described must be understood to represent the majority of specimens existing in their preferred habitats. Specimens surviving in stressed conditions (too much or too little water or sun) may produce stunted or discolored flowers. If *Spiranthes* is examined on a hemisphere-wide basis, it becomes clear that many species such as *S. cernua* and *S. ochroleuca* exhibit a complex collection of physical variations that make species identification confusing. Orchidologists such as Charles J. Sheviak (1982) in New York have compiled some fascinating studies on the range and occurrence of these variations. The limited geographic range associated with this guide allows us to deftly sidestep many of these perplexing classification issues, but keep in mind that attempting to apply these keys to orchids in other areas may lead to misleading results.

It has been traditional in orchid literature to base *Spiranthes* keys on whether or not the flowers are single-ranked or multi-ranked on the stem, and this work will continue that tradition. Even here we must apply caution in relying only on this one characteristic, for young or environmentally stressed specimens of a usually multi-ranked species will occasionally produce singly-ranked flower spikes. Conversely, single-rank species will occasionally produce multi-ranked specimens, as if to keep us from being too sure of ourselves. To minimize the possibility of error, each species in the key will also have several other associated or confirming characteristics included. If any of these additional traits are not present, it may be necessary to study the full descriptions to make a final determination.

One final word of caution: these descriptions reflect the Spiranthes species as they are currently understood. If the reader tries to match these descriptions to those plants listed under identical names in orchid texts of even a few years ago some contradictions in characteristics and range will be noted. This occurs when an "old" species is split into two "new" species, or when what was once thought to be two species is later believed to represent only local varieties of a single species. So with these warnings and cautions in mind, let us begin our in-depth look at *Spiranthes,* keeping in mind that these facts may be open to reinterpretation in the future!

Key to Spiranthes*
(continued from page 8)

I. Flowers usually single-ranked on the stem. (Single-ranked flowers present as a loose spiral of blooms that appear to be evenly spiraled around the stem when viewed from above.)
   A. Lip white, tiny—about 3 mm (1/8″) long.
      1. *S. tuberosa* → page 71.
         Additional information: leaves broad, but usually not present during flowering. Flowers August. Range in northeast limited to southeastern New York, Massachusetts, Connecticut, Rhode Island.
   B. Lip white with a green center, about 5 mm (1/5″) long.
      2. *S. lacera* → page 73.
         Additional information: leaves broad, may or may not be present during flowering. Flowers July and August. Range throughout the northeast.
   C. Lip white to creamy white with a yellowish center, 5–8 mm (1/5–3/10″) long.
      3. Inflorescence covered with minute light-colored pointed hairs. Lip 6–8 mm (1/4–3/10″) long.
         S. vernalis → page 75.
         Additional information: northeast range includes southeastern New York, Massachusetts, Connecticut, Rhode Island. Flowers late July into August.
      4. Inflorescence covered with minute reddish brown round-headed hairs. Lip 5–7 mm (1/5–1/4″) long.
         S. casei → page 77.
         Additional information: northeast range includes Maine, Vermont, New Hampshire, northeastern New York. Flowers late August to mid-September.
II. Flowers usually multi-ranked on the stem. (Multi-ranked flowers present as very densely flowered spirals, flowers appear to be lined up into 3 or 4 vertical stacks or ranks when viewed from above.
   A. Lip oblong, bright yellow with white margins, about 5 mm (1/5″) long. Leaves lance-shaped.
      5. *S. lucida* → page 79.
         Additional information: most flower in June and July. Usually the first *Spiranthes* to bloom in an area.

B. Lip egg-shaped with a shallow constriction about the middle, white to yellowish white to white with a yellowish center, about 10 mm (²/₅″) long.
  6. Leaves long and slender with short or no stalk, growing from base and up onto stem. Lip typically white to yellowish white.
  S. ochroleuca → page 81.
  Additional information: prefers drier shady habitats than *S. cernua*. Flowers September into October.
  7. Leaves long and slender with a distinct base, most growing from the base of the stem. Lip typically white to white with a yellowish center.
  S. cernua → page 83.
  Additional information: prefers open moist habitats. Flowers late August to September.
C. Lip deeply constricted about the middle, white to creamy white, about 10 mm (²/₅″) long.
  8. *S. romanzoffiana* → page 85.
  Additional information: Flowers July and August.

* This key and the following descriptions incorporate information from *The Biosystemic Study of the Spiranthes cernua Complex* by Sheviak (1982) and *Spiranthes casei, A New Species from Northeastern North America* by Catling and Cruise (1974).

Spiranthes tuberosa Rafinesque

Etymology: *tuberosa* = referring to the characteristic single tuber of this species.

Common names: little ladies' tresses, Beck's tresses, little pearl-twist.

Description: Leaves 2–3, 5–6 cm (2–2⅖″) long, egg-shaped to oblong, stalk about ⅖ total leaf length, growing from the base of the stem, green, usually not present during flowering. **Stem** 13–22 or up to 30 cm (5–9 or up to 12″) tall (reports of much taller specimens in some guides may reflect either mistaken identifications or atypically large specimens), slender, smooth, with several sheathing bracts, green. **Inflorescence** about ⅕–¼ of the total stem, usually 2.5–5 cm (1–2″) long, with 15–20 or up to 30 flowers, flowers loosely to densely arranged in a single rank. **Flowers** about 2 mm (1/12″) tall and wide, 3 mm (⅛″) long, tubular, all parts white. **Sepals** and **petals** similar, about 3 mm (⅛″) long, narrowly lance-shaped; dorsal sepal and petals converge to form a sheltering hood over the lip, tips of all three parts slightly upturned, lateral sepals spread slightly outward and down; **lip** about 3 mm (⅛″) long, egg-shaped with a wavy and unevenly ragged outer margin, tip somewhat downturned.

Habitat: usually in dry and sandy to rocky soils in meadows and open woodlands, occasionally in moister areas.

Flowering season: Most bloom in August, some flower to the second week of September.

Range: all of the southeastern quarter of the United States (excluding only southern Florida) north of Michigan and Massachusetts. In the northeast it is found in southeastern New York, Massachusetts, Connecticut, and Rhode Island.

Comments: Living up to its name, the little ladies' tresses does indeed have the smallest flowers of any *Spiranthes* in the northeast. *Spiranthes* as a group has been notoriously difficult to classify into distinct species, with the result that the older nomenclature can be quite confusing. This difficulty peaked with the obsolete species name of *beckii,* which has been applied at different times to *Spiranthes tuberosa, S. lacera* var. *gracilis,* and *S. praecox!*

Spiranthes lacera (Rafinesque) Rafinesque var. **lacera**
Spiranthes lacera var. **gracilis** (Bigelow) Luer

Etymology: *lacera* = "torn," as in "lacerated," referring to the ragged lip margin; *gracilis* = "slender," referring to the stem.

Common name: slender ladies' tresses.

Description: var. **lacera: Leaves** 3–5, 2–5 cm (⁴/₅–2″) long, egg-shaped, short stalked, basal, present during flowering. **Stem** 20–50 cm (8–20″) tall, somewhat pubescent, with several sheathing bracts, green. **Inflorescence** ¹/₄–¹/₃ or more of the total stem, usually 5–20 cm (2–8″) long, with 30–40 flowers, flowers loosely arranged in a single rank, loosely spiraled or occasionally one-sided on the stem. **Flowers** about 4 mm (¹/₆″) tall and wide, 5 mm (¹/₅″) long, tubular, white with a greenish center, fragrant. **Sepals** and **petals** similar, 4–5 mm (¹/₆–¹/₅″) long, narrowly lance-shaped, white; dorsal sepal and petals converging to form a tubular hood over the lip, the tips of all three parts curving upward, lateral sepals spread slightly outward and down; **lip** about 5 mm (⅕″) long, oblong with a raggedly toothed outer margin, tip downturned, greenish with wide white margin.

Description: var. **gracilis:** as above, with the following differences: leaves usually not present at flowering; stem more slender and usually smooth throughout; inflorescence closely arranged in a tight spiral that curves around the stem several times; height of stem up to 60 cm (24″) tall and flower parts slightly longer than in the typical form (add 1–2 mm per part). Flowers occasionally arranged in more than one rank.

Habitat: sandy soils in dry meadows or woodland clearings, often in previously disturbed areas such as roadsides.

Flowering season: second week of July to fourth week of August, a few into September (third week of July to second week of August). Luer (1975) noted that *S.* var. *lacera* blooms peak before *S.* var. *gracilis*. There is however great overlap between the two, and in those locations where they exist side by side it is possible to photograph both in bloom on the same day.

Range: *Spiranthes lacera* var. *lacera* is found throughout the northeast quarter of the United States north into southeastern Canada, while var. *gracilis* is found throughout the eastern half of the United States except for most of Florida and Maine, and northern New Hampshire.

Comments: Those two interesting species of human beings known as "splitters" and "lumpers" hotly debated the true taxonomic status of the slender ladies' tresses for years. After the dust cleared, most (but not all) of the splitters licked their lumps and agreed to accept *S. lacera* as a single species with two distinct varieties, not the two separate species they originally believed to exist. This gentleman's agreement is easy to accept in the extremes of this species' range, for in the southeast var. *gracilis* stands supreme, while var. *lacera* rules from Nova Scotia to Lake Winnipeg. Where the two varieties overlap, all of the northeast save for Maine, the situation is more confusing. I have seen stands of this orchid where the variety in question was easy to ascertain, but I also know of one location where both varieties and every possible combination of characteristics between the two exist within one population.

Spiranthes vernalis Engelmann and Gray

Etymology: *vernalis* = "spring," the flowering time of this orchid in the southeastern United States.

Common name: spring ladies' tresses.

Description: Leaves 4–5, 5–15 or up to 25 cm (2–6 or up to nearly 10″) long, long and slender, growing from the base or lower portions of the stem, green, present during flowering. **Stem** 15–60 cm (6–24″) tall, with minute light-colored and pointed hairs on the upper portion, with several sheathing bracts, green. **Inflorescence** ¼–⅓ of the total stem, usually 5–13 cm (2–5″) long, with 25–50 flowers, flowers usually densely arranged in a single rank, spiraled or occasionally one-sided on the stem. **Flowers** about 6 mm (¼″) tall and wide, 8 mm (³⁄₁₀″) long, tubular, white to off-white with a yellowish center. **Sepals** 6–9 mm (¼–⅓″) long, narrowly lance-shaped, minutely pubescent, white; **petals** 6–9 mm (¼–⅓″) long, narrowly oblong, white; dorsal sepals and petals converging to form a tubular hood over the lip, the tips of all three parts curving upward, lateral sepals spread slightly outward and down; **lip** 6–8 mm ($1/4$–$3/10$″) long, egg-shaped with a wavy and raggedly uneven outer margin, tip downturned, yellowish with a wide off-white margin.

Habitat: usually damp to moist soils from open meadows and fens to pine barrens, occasionally in dryer situations.

Flowering season: third week of July to second week of September, usually peaking in early to mid-August.

Range: the southeastern United States west to Texas and Oklahoma. Along the east coast it ranges north to southeastern New York, Massachusetts, Connecticut, and Rhode Island.

Comments: A rather tall ladies' tresses, often associated with coastal areas. Under a hand lens, the flowers are revealed as one of the "hairiest" of ladies' tresses blooms in the northeast. The common name of spring ladies' tresses is not very accurate in the northeast, where blooming does not occur until late summer.

Spiranthes casei Catling and Cruise

Etymology: *casei:* in honor of Frederick W. Case, author of *Orchids of the Western Great Lakes Region* (1964). Case recognized that these orchids, which now bear his name, were probably not properly classified at that time.

Common names: Case's ladies' tresses, northern vernalis.

Description: Leaves 2–4, 5–20 cm (2–nearly 8″) long, narrowly lance-like with a slender stalk, growing from the base or the lower portions of the stem, green, the basal leaves often absent at flowering, green. **Stem** 25–35 or up to 43 cm (10–14 or up to nearly 17″) tall, smooth below but with minute reddish-brown round-headed hairs on the upper portions, with several sheathing bracts, green. **Inflorescence** about ¼ of the total stem, usually 6–10 up to 15 cm (2⅖–4 up to 6″) long, with 15–30 flowers, flowers usually loosely arranged in a single spiraled rank. **Flowers** about 4–5 mm (⅙ to ⅕″) tall and wide, 7 mm (¼″) long, tubular, white to off-white to creamy white. **Sepals** and **petals** similar, 5–7 mm (1/5–1/4″) long, narrowly lance-shaped, minutely pubescent, off-white to creamy white, dorsal sepal and petals converging to form a tubular hood over the lip, tips of all three parts pointing forward, lateral sepals spread slightly outward; **lip** 5–7 mm (1/5–1/4″) long, egg-shaped with a wavy and a raggedly uneven outer margin, tip down-curved, white to off-white to creamy white with a yellowish base.

Habitat: usually found in previously disturbed dry or well-drained sandy soils in open locations such as roadsides and old fields.

Flowering season: fourth week of August to second week of September (fourth week of August to first week of September).

Range: from Nova Scotia across southern Canada and northern New England to upper Michigan. In the northeast it is found in Maine, northern Vermont and New Hampshire, and northeastern New York.

Comments: Case's ladies' tresses is very similar to *S. vernalis,* and the two were until recently believed to be the same species. Under a hand lens the upper stems of both species are revealed to be covered with fine hairs, but those of *S. vernalis* are lighter colored and pointed, while the hairs on *S. casei* are reddish brown with rounded heads.

Spiranthes lucida (H. H. Eaton) Ames

Etymology: *lucida* = "shining," referring to the leaves. In astronomy, a lucida is the brightest star in a group, making this species the brightest star in the constellation *Spiranthes*.

Common names: wide-leaved ladies' tresses, shining ladies' tresses.

Description: Leaves 3–4, 2.5–12 cm (1–4½″) long, lance–shaped, the base of the leaves sheathing the stalk, growing from the base or lower portions of the stem, glossy green, present during flowering. **Stem** 10–25 or up to 35 cm (4–10 or up to nearly 14″) tall, smooth below to slightly pubescent above, with a few sheathing bracts, green. **Inflorescence** about ⅕–¼ of the total stem, usually 2.5–7.5 or up to 11 cm (1–3 or up to 4³⁄₁₀″) long, with 10–20 flowers densely arranged in three ranks. **Flowers** about 4 mm (⅙″) tall and wide, 5–6 mm (⅕–¼″) long, tubular, white with a brilliant yellow throat. **Sepals** and **petals** similar, about 5 mm (⅕″) long, somewhat oblong, white; dorsal sepal and petals converge to form a sheltering hood over the lip, their tips pointing forward or slightly upcurved, lateral sepals pressed against the petals, their tips curing slightly outward and downward; **lip** about 5 mm (1/5") long, oblong with a wavy and unevenly toothed outer margin, downturned at the tip, bright yellow with white margins.

Habitat: moist and often previously disturbed soils on grassy shores of waterways and open swampy meadows.

Flowering season: first week of June to fourth week of July, a few reported earlier and later than this (fourth week of June to second or third week of July).

Range: Nova Scotia to New Jersey, westward to Indiana and Kentucky. It is found throughout the northeast except in northern Maine and eastern Massachusetts.

Comments: In most localities *Spiranthes lucida* is the first of the ladies tresses to flower each year. The early blooming season, bright yellow-centered lip, small size and glossy oval leaves combine to make this one of the easiest ladies' tresses to identify.

Spiranthes ochroleuca (Rydberg) Rydberg

Etymology: *ochroleuca: ochro* = "pale yellow," *leuca* = "white."

Common name: yellow nodding ladies' tresses.

Description: Leaves 3–6, 5–20 cm (2–nearly 8″) long, long and slender with either a very short stalk or no stalk at all, growing from the base of and up onto the stem, green, present during flowering. **Stem** 15–30 or up to 50 cm (6–12 or up to 20″) tall, pubescent on the upper portions, with several sheathing bracts, green. **Inflorescence** ¼–⅓ of the total stem, usually 7.5–12 cm (3–4½″) long, with 15–30 or up to 60 flowers, flowers usually densely arranged and multiranked. **Flowers** about 5–8 mm (³⁄₁₀″) tall and wide, 10–12 mm (⅖″) long, tubular, all parts white to pale yellowish white. **Sepals** and **petals** similar, 10–11 mm (⅖″) long, narrowly lance-like; dorsal sepal and petals converging to form a tubular hood over the lip, the tips of all three parts curved upward, lateral sepals pressed against the petals, tips usually pointing forward; **lip** about 10–11 mm (²⁄₅″) long, egg-shaped and slightly constricted near the center with a wavy and usually smooth margin, tip strongly downcurved or recurved.

Habitat: dry or well drained soils, especially in woodland clearings and edges.

Flowering season: fourth week of August to first week of October (second to fourth week of September).

Range: from Newfoundland and Nova Scotia south to northern New Jersey and westward to the middle of the Great Lakes region. The yellow nodding ladies' tresses is found throughout the northeast except for northern Maine.

Comments: As the common name of yellow nodding ladies' tresses suggests, this species is very similar to the true nodding ladies' tresses. They are so similar that at one time they were believed to be two varieties of a single species. When compared, the yellow nodding ladies' tresses are often dingier in color (but nowhere near true yellow), prefer drier and shadier habitats, and are slightly later blooming in the northeast. The yellow nodding ladies' tresses is probably the latest blooming orchid in New York and New England.

Spiranthes cernua (Linnaeus) L. C. Richard

Etymology: *cernua* = "nodding," supposedly in reference to a slightly drooping position of the blooms on the stem, but in my experience this lowered posturing occurs more often in literature than in nature!

Common names: nodding ladies' tresses, autumn ladies' tresses.

Description: Leaves 3–6, 5–20 cm (2–nearly 8″) long, long and slender with a distinct stalk, most growing from the base of the stem, green, present during flowering. **Stem** 15–38 or up to 50 cm (6–15 or up to 20″) tall, pubescent on the upper portions, with several sheathing bracts, green. **Inflorescence** $1/4$–$1/3$ of the total stem, usually 7.5–10 cm (3–4″) long, with 15–30 or up to 60 flowers, flowers usually densely arranged and three-ranked. **Flowers** about 7–8 mm ($1/4$–$3/10$″) tall and wide, 10 mm ($2/5$″) long, tubular, white often with a yellowish or greenish center. **Sepals** and **petals** similar, about 10 mm

(²⁄₅″) long, narrowly lance-shaped, white; dorsal sepal and petals converging to form a tubular hood over the lip, tips of all three parts curved slightly upward, lateral sepals pressed against the petals, tips curved slightly inward; **lip** about 10 mm (²⁄₅″) long, egg-shaped and slightly constricted near the center with a wavy and usually minutely toothed outer margin, tip strongly downturned or recurved, white with yellowish or greenish near the base.

Habitat: open moist areas, such as roadside ditches, the edges of swampy fields, springs, and fens.

Flowering season: third week of August to first week of October, a few reported earlier and later than this (first to third week of September).

Range: The nodding ladies' tresses orchid is found throughout the eastern half of the United States, except for southern Florida and northernmost Maine. It is also found in the southeastern tip of Canada. It is found throughout the northeast except for northern Maine.

Comments: The nodding is the most commonly encountered ladies' tresses in the northeast. In early autumn large numbers may be found in a wide variety of moist to wet situations.

Spiranthes romanzoffiana Chamisso

Etymology: *romanzoffiana:* in honor of Nicholas Romanzoff, a Russian leader and patron of science who lived from 1754–1826.

Common name: hooded ladies' tresses.

Description: Leaves 3–6, 7.5–20 or up to 25 cm (3–8 or up to nearly 10″) long, long and slender, growing from the base of and up onto the stem, green, present during flowering. **Stem** 15–30 or up to 50 cm (6–15 or up to 20″) tall, smooth below, becoming pubescent on the upper portions, with several sheathing bracts, green. **Inflorescence** about ⅓ of the total stem, usually 5–12 cm (2–4½″) long, with 15–50 or up to 60 flowers, flowers densely arranged in three ranks on the stem. **Flowers** about 8–9 mm (³⁄₁₀–⅖″) tall, 4–5 mm (⅙–⅕″) wide, 9–11 mm (⅓–⅖″) long, tubular with a nearly spherical or bulbous base when viewed from the side, all parts white to creamy white, fragrant like almonds (Correll 1950) or courmarin (Sheviak 1982) or violets (Baldwin 1884). **Sepals** and **petals** similar, usually 9–11 mm (⅓–⅖″) long, lance-like to narrowly lance-like, all sepals and petals join to form a tubular hood over the lip, hood angled out from the stem at about 45° (flowers of most similar northeastern multi-ranked *Spiranthes* angle out from the stem at about 90°); **lip** 9–10 mm (⅓–⅖″) long, deeply constricted about the middle so as to appear fiddle-shaped, outer margin minutely toothed, tip downcurved and recurved from a point only halfway up the hood.

Habitat: highly variable, usually but not always in moist to wet areas, often in strong sunlight but also frequently in semishaded locations.

Flowering season: third week of July to first week of September (fourth week of July to fourth week of August).

Range: widespread, from Alaska across central Canada south to New York and Ohio in the east and Arizona and New Mexico in the west. It is also found in Ireland and nearby islands. In the northeast it is found throughout except for southeastern New York, Connecticut, Rhode Island, and eastern Massachusetts.

Comments: The hooded ladies' tresses is very commonly encountered in fens during the month of August. The bulbous base of the flower and the flower's upward angle on the stem, combined with the deeply constricted fiddle shaped lip, make this species easy to identify.

Group 5

Tipularia
Aplectrum
Corallorhiza
Triphora
Galearis
Amerorchis
Epipactis
Coeloglossum
Platanthera
Malaxis
Listera
Liparis

Tipularia

Tipularia discolor (Pursh) Nuttall

Etymology: *Tipularia:* referring to the insect genus *Tipula,* the craneflies, which the flowers resemble; *discolor* = "mottled, blotched," which likely refers to the leaves but could also be used to describe the flowers.

Common name: cranefly orchid.

Description: Leaf solitary, 5–7.5 or up to 10 cm (2–3 or up to 4″) long, stalk up to 5 cm (2″) long, broadly egg-shaped with a pointed tip, dark green frequently with some purplish mottling above, purple on the undersurface. **Stem** 30–50 or up to 60 cm (12–20 or up to 24″) tall, wrapped in a leaf-like sheath near the base, brownish. **Inflorescence** about ½ or more of the stem, usually 10–25 cm (4–10″) long, with 20–40 loosely arranged flowers. **Flowers** about 13 mm (½″) tall and wide, color of all parts pale and variable, ranging from yellowish brown to green to lavender or purple, sometimes with purplish mottling. **Sepals** and **petals** similar, about 6 mm (¼″) long, narrowly lance-shaped, spreading outward from the center; **lip** about 7 mm (¼″) long, 3-lobed, lateral lobes small, central lobe long and spreading slightly at the tip, with a long (about 20 mm [⅘″]) usually horizontal **spur**.

Habitat: shady and sometimes somewhat damp hardwood forests with a rich layer of humus.

Flowering season: fourth week of July to second week of August, rarely into September.

Range: The cranefly orchid inhabits most of the southern and mid-Atlantic states west to the Mississippi River, excluding much of Pennsylvania and Florida. In the northeast its present day range is limited to Long Island and Massachusetts.

Comments: Few if any orchids in the United States look less like an orchid to the casual observer than the cranefly orchid. At first glance the flowering stem of pale purple flowers appears fairly ordinary, perhaps reminiscent of the lily family. It is only on close examination that the telltale column and long-spurred lip reveal the true nature of this species. Close examination reveals that the blooms have a somewhat asymmetrical appearance, a characteristic resulting from each flower being tipped on its side by a twisted base support. In this position each bloom truly lives up to the common name of cranefly orchid, for the long, horizontally directed spur does resemble the elongated abdomen of the true craneflies, and it takes only a little imagination to turn the half-inch petals and sepals into the wings and legs of these common insects.

At first glance the cranefly may not look like an orchid, but its natural history is in many ways parallel to that of another woodland orchid, the putty-root. Both species grow from a series of corms about 2 cm (⅘″) long. With the cranefly orchid there may be from 2–6 of these rootbearing structures, each directly connected to the next, like pearls strung on a necklace. Each year in mid-September the youngest of the corms puts up a single dark green to purplish leaf, hence the old

Latin name of *T. unifolia*, "one leaf." This single rather tough leaf survives all but the most brutal winter weather, only to wither and die the following spring, just as most other plants are displaying their first new year greenery. Nothing more is seen from the cranefly until midsummer, when the flowering stem finally appears. It does not take too great a leap of imagination to envision each loose cluster of flowers as a cloud of diminutive craneflies circling together about a foot above the forest floor. The flowers, which seem to peak in early August in the northeast, last long enough that it is not unusual to find a stalk with all the flowers in bloom at the same time. It is interesting to note that only a small percentage of the *Tipularia*s bloom each year, so finding a large number of leaves is no indication of how many flowers will be found six months hence.

Aplectrum

Aplectrum hyemale (Muhlenburg ex Willdenow) Nuttall

Etymology: *Aplectrum* = "spurless" (many multiflowered orchids sport spurred lips); *hyemale* = "winter," referring to the leaf.

Common names: putty-root, Adam-and-Eve, winter-leaf.

Description: Leaf solitary, 10–15 or up to 20 cm (4–6 or up to nearly 8″) long, broadly oval with a pointed tip, strongly pleated with parallel veins, short-stalked, silvery green above with purplish tints on the underside. **Stem** 25–40 or up to 50 cm (10–16 or up to 20″) tall, wrapped in 2–3 tall tubular leaflike sheaths, pale green. **Inflorescence** about ⅕–¼ of stem, usually 5–10 cm (2–4″) long, with 8–12 or up to 15 flowers. **Flowers** about 19 mm (¾″) tall, 25 mm (1″) wide, and 12 mm (⅖″) long, pale yellowish green to purplish green. **Sepals** and **petals** similar, about 13 mm (½″) long, narrowly lance-shaped, dorsal and lateral sepals spreading, petals growing forward above the lip, off-white to yellowish white to pale green highlighted with purplish tints; **lip** about 10–12 mm ($2/5''$) long, 3-lobed, lateral lobes small and erect, central lobe large and fan-shaped, ridged near the base, margin uneven and wavy, white with a few delicate purplish spots and veins. In forma *pallidum* the lip is white with no purplish markings.

Habitat: old and established open stands of deciduous trees, especially beech and maple, with a rich layer of leaf humus underneath.

Flowering season: fourth week of May to second week of June (first to second week of June).

Range: Vermont and southern Canada south to North Carolina, west to the Mississippi River basin and the southern Great Lakes. In the northeast putty-root is now found in eastern New York, western Vermont, and Massachusetts.

Comments: Both common names of this interesting orchid refer to the roots. Putty-root refers to a natural glue that is said to be obtained by crushing the corms. It is reputed to be strong enough to mend glassware, but it would be a crime against nature to put a protected plant to such a poor use! The name Adam and Eve refers to the fact that the underground portion of this plant usually consists of two circular corms about 25 mm ($1''$) across. The two corms are joined by a short strand, hence the reference to the original couple Adam and Eve. This twin corm system is the central element in the unusual natural history of this orchid. In early September the youngest corm in the pair forms a new corm which puts up the single leaf previously described. This hardy blade persists through the brutal winter weather, only to wither away the following spring when most other woodland residents are just displaying their earliest greenery. As this newest corm developed, the oldest corm in the chain passed away. Sometime between mid-May and mid-June the same young corm which produced the leaf may send up a single multi-flowered stem, after which it enjoys a well-deserved rest until late summer, when it will produce the next corm in the chain. It is interesting to note that only a small percentage of leaf-bearing corms produce a flowering stem each year. With its multiflowered stalk and no leaves in evidence at flowering time, the putty-root was once believed to be a member of *Corallorhiza,* a group of leafless orchids.

Corallorhiza
The Coralroots

Corallorhiza: coral = "coral-like," *rhiza* = "root."

The genus *Corallorhiza* is known to have ten members, of which six are found in the United States. Four of these inhabit the northeast. Coralroots are well named, for their multi-branched underground rhizomes do in fact have a strong resemblance to some types of marine corals. The showy floral parts of these orchids grow from the stout tip of the combined pedicel/ovary, which is typically as long as the flowers themselves, giving the visual impression of a beautiful diminutive orchid growing from a long, fat "stalk." Following pollination the flowers wilt and the ovaries swell into drooping oval seed capsules.

Coralroots are seemingly leafless orchids whose unusual mode for obtaining nutrients has created some confusion about their true nature. Coralroots were originally thought to be direct parasites, vegetative vampires who stole their sustenance from the unprotected roots of host plants. This belief led to the creation of some pretty colorful folk tales and also some fairly lurid fiction from writers such as H. G. Wells. As the investigations continued, coralroots were later thought to be saprophytes, organisms that obtain their nutrients from dead and decaying organic matter in the soil. It was noted that, like other orchids, coralroots share a symbiotic lifestyle with a subterranean fungus. At one time it was believed that these fungi were decomposing only dead or decaying plant matter. This fungus processes available nutrients into a form easily absorbed by the orchid. It is now thought that these fungal intermediates feed directly on other seed plants, leading to the current view of coralroots as being indirectly parasitic.

Typical flowering plants manufacture needed nutrients by exposing chlorophyll-laden leaves to sunlight. Because *Corallorhiza* are essentially nonphotosynthetic, these structures reduced to a few tubular sheaths on the flowering stem. Different coralroots do contain viable amounts of chlorophyll, but in only one species, *C. trifida*, are the amounts large enough and unmasked by other pigments to color this orchid green. The remaining species are free to take their colors from other parts of the visual spectrum, and are painted in all shades of yellows, tans, pinks, and purples. Since coralroots are not dependent on photosynthetic activity to survive, these orchids can exist unobserved for years below the surface. Only on years when conditions are appropriate to justify the reproductive effort will the coralroot send a flowering stem skyward, the only above-ground proof of its subterranean existence.

Key to Corallorhiza
(continued from page 9)

I. Lip with 3 lobes
   1. Lip 6–7 mm (¼″) long. White with purple spots. Flowers July and August.
      C. maculata → page 96.
   2. Lip 4–5 mm (⅙–⅕″) long. White with or without a few purplish spots near the base. Flowers May and June.
      C. trifida → page 98.
II. Lip without lobes (somewhat egg-shaped to tongue-shaped).
   3. Lip 10–12 mm (⅖″) long. Tongue-shaped with smooth margins. Lip somewhat striped or appearing solidly colored. Sepals and lateral petals distinctly striped. Flowers June.
      C. striata → page 100.
   4. Lip egg-shaped with a wavy margin. White with purple spots. Lip about 4 mm (⅙″) long. Flowers August and September.
      C. odontorhiza → page 102.

Corallorhiza maculata (Rafinesque) Rafinesque var. **maculata**

Etymology: *maculata* = "spotted," referring to the lip.

Common name: spotted coralroot, large coralroot.

Description: Leaves lacking. **Stem** 20–48 or up to 80 cm (8–19 or up to just over 30″) tall, smooth, with several leaf-like sheaths, purplish to pale brown. **Inflorescence** $1/3$–$1/2$ of the stem, usually 5–20 cm (2–8″) long, with 10–30 or up to 40 loosely arranged flowers. **Flowers** about 12 mm ($2/5''$) tall and wide, 7 mm ($1/4''$) long, greenish purple and white with purple spots, originally growing upward from the stem, later drooping following pollination. **Sepals** about 8 mm ($3/10''$) long, narrowly lance-shaped, usually purplish green with 3 more or less distinct purplish veins; **petals** 6–7 mm ($1/4''$) long, narrowly lance-shaped, colored like the sepals or sometimes lighter; sepals and petals spreading outward and somewhat forward from the center; **lip** 6–7 mm ($1/4''$) long, somewhat egg-shaped, 3-lobed, white with purple spots and markings, side lobes small and erect, central lobe broad with a wavy margin, somewhat downcurved.

Habitat: the spotted coralroot grows in rich humus in hardwood, coniferous, or mixed woods.

Flowering season: fourth week of June to fourth week of August or occasionally later, but see Comments below (fourth week of July to third week of August).

Range: from North Carolina north through all of the northeast up to Newfoundland, westward around the Great Lakes through southern Canada to the Pacific, then south through the coastal states to mid-California and south along the Rocky Mountains to Arizona and New Mexico. There are also large populations in southern Mexico and Guatemala.

Comments: *Corallorhiza maculata* is probably the most commonly encountered coralroot in the northeast. In a number of earlier botanical texts Wister's coralroot, *C. wisteriana*, was listed as being found in Massachusetts. I have not been able to locate any populations of that species in the northeast and am now of the belief that those reports may have been based on incorrectly identified *C. maculata*. These two species are superficially similar, but Wister's is smaller, has an unlobed lip, less spreading sepals, and an earlier flowering season. Paul Martin Brown has shown me a stand of the western spotted coralroot, *C. maculata* var. *occidentalis* in New England. He reports that they bloom from mid-June through mid-July, a full month in advance of the commoner variety described above. This is pure conjecture, but perhaps it was one of these earlier blooming orchids that was responsible for a mistaken reporting of the also equally early flowering Wister's coralroot.

Corallorhiza trifida Chatelain

Etymology: *trifida* = "in three parts," referring to the three lobes of the lip.

Common names: early coralroot, northern coralroot.

Description: Leaves lacking. **Stem** 7.5–30 cm (3–12″) tall, slender, smooth, with 2–5 leaf-like sheaths, pale yellow to yellowish green to green. **Inflorescence** ¼–⅓ of the stem, usually 2.5–7.5 cm (1–3″) long, with 10–15 or up to 20 loosely arranged flowers. **Flowers** about 6–9 mm (¼–⅓″) tall, 5–8 mm (⅕–³⁄₁₀″) wide, 5 mm (⅕″) long, usually greenish to greenish white and sometimes tinged with purplish brown, growing upward or outward from the stem, later drooping following pollination. Dorsal **sepal** 5–6 mm (⅕–¼″) long, narrowly lance-shaped; lateral sepals 5–6 mm (⅕–¼″) long, narrowly sickle-shaped; **petals** 4–5 mm (⅙–⅕″) long, narrowly lance-shaped; sepals and petals greenish, sometimes tinged with purplish brown near the tips; dorsal sepal and petals bend forward to form a loose hood over the lip, lateral sepals spread slightly outward and downward; **lip** about 4–5 mm (1/6–1/5″) long, wedge-shaped to somewhat egg-shaped, 3-lobed, lateral lobes small and erect, central lobe broad with a wavy margin, downcurved, white with or without a few purplish spots near the base.

Habitat: rich or moist forests.

Flowering season: third week of May to fourth week of June, rarely early July (second to fourth week of June).

Range: most of Canada and Alaska, south along the Rocky Mountains to Arizona in the west and south through Pennsylvania in the east. It is found throughout the northeast. It is also found throughout most of Europe and much of northern Asia.

Comments: The coralroots are well known for their apparent lack of green pigmentation. This makes the early coralroot the oddball of this orchid group, for it is typically pale yellowish-green in color.

Corallorhiza striata Lindley var. **striata**

Etymology: *striata* = "striped," referring to the flowers.

Common names: striped coralroot, Macrae's coralroot.

Description: Leaves lacking. **Stem** 15–50 cm (6–20″) tall, stout, smooth, enclosed by leaf-like sheaths, yellowish to more often purplish. **Inflorescence** up to ⅔ of the stem, usually 5–20 cm (2–8″) long, with 15–25 or up to 35 loosely arranged flowers. **Flowers** about 20 mm (⅘″) high, 15–25 mm (⅗–1″) wide, 20 mm (⅘″) long, pinkish to purplish, usually somewhat drooping. **Sepals** and **petals** similar, about 10–15 mm (⅖–⅗″) long, somewhat lance-shaped, pale pink to yellowish pink with 3–5 conspicuous purplish veins, all spreading outward and somewhat forward from the center; **lip** about 10–12 mm (⅖″) long, tongue-shaped with a smooth margin, white with 5 wide reddish to purplish veins, the stripes sometimes so wide as to make the entire lip purplish.

Habitat: variable; from dry to somewhat moist soils in hardwoods, conifers or mixed woods, often associated with limestone soils. The only known population in the area included in this book exists in a wooded swamp in New York.

Flowering season: second and third weeks of June.

Range: The striped coralroot is found in a band across southern Canada from the Atlantic to the Pacific. It ranges south along the Rocky Mountains into Wyoming and almost halfway down California. Isolated populations of *C. striata* occur in suitable habitats into southern Mexico. The northeastern United States is usually considered south of this species' range in the east, but one station occurs in New York.

Comments: "If you haven't seen it in full sunlight, you really haven't seen the striped coralroot." This observation is one of the commonest applied to this species, for only in bright light is the jewel-like nature of its pinkish purple colorations fully revealed. The striped coralroot may be the rarest member of this group in the northeast, but its conspicuously striped floral parts make it by far the easiest species to identify. Variety *striata* differentiates the typical form described here from var. *vreelandii*, a smaller and paler western variety with a limited range.

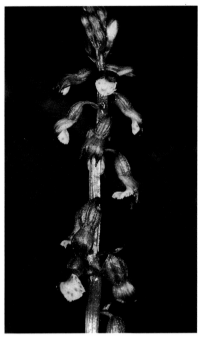

Courtesy George H. Beatty

Corallorhiza odontorhiza (Willdenow) Nuttall

Etymology: *odontorhiza: odonto* = "tooth," *rhiza* = "root."

Common names: autumn coralroot, small coralroot.

Description: **Leaves** lacking. **Stem** 13–20 but reported up to 40 cm (5–8 and up to 16″) tall, very slender and bulb-like at the base, smooth, with several leaf-like sheaths, straw-colored to purplish. **Inflorescence** about ¼ of the stem, usually 5–7.5 cm (2–3″) long, with 5–15 or reportedly up to 20 flowers. **Flowers** about 2.5–3 mm (¹⁄₁₀–⅛″) tall and wide, 5 mm (⅕″) long, greenish purple and white with purple spots, flowers angled out from stems at just over 90° or drooping. **Sepals** and **petals** similar, about 4 mm (⅙″) long, lance-shaped, a mixture of green and brownish-purple, all growing forward over the lip giving the flower an almost tubular appearance; **lip** about 4 mm (⅙″) long, egg-shaped with a wavy margin, white with purple spots and markings.

Habitat: The autumn coralroot grows from humus or often damp sandy soils in hardwoods, conifers, or mixed woods.

Flowering season: fourth week of August to fourth week of September, a few reported both earlier and later than this (first to third week of September).

Range: most of the eastern half of the United States, excluding much of Florida, northern New York, Vermont and New Hampshire, and most of Maine. It is also found in southern Mexico and several Central American countries.

Comments: The autumn coralroot is easily identified by its tiny, barely opening flowers and its swollen bead-like ovary. Usually overlooked because of its small size, under a hand lens the bloom displays a delicate beauty.

Key to Orchids with Several to Many Small to Medium Size Flowers on a Single Stem *(continued from page 9)*

1. Lips with a prominent elongated and slender spur.
    A. Plant typically with one leaf at flowering time.
        1. Flowers small but showy, white with pinkish spots. *Amerorchis rotundifolia* → page 110.
        2. Flowers small, greenish, inconspicuous. *Platanthera* → page 120.
    B. Plant typically with two leaves at flowering time. Leaves growing from or near the base of the stem.
        1. Flowers bicolored, white and pale purplish. *Galearis spectabilis* → page 108.
        2. Flowers off-white, greenish white or yellowish green throughout. *Platanthera* → page 120.
    C. Plant typically with several leaves at flowering time, leaves arranged alternately on an upright stem. *Platanthera* → page 120.

II. Lips not having a slender elongated spur, either having no spur or having a rounded saccate spur.
    A. Plants typically with one leaf at flowering time. *Malaxis* → page 163.
    B. Plants typically with two leaves at flowering time.
        1. Leaves opposite, growing from a point about halfway up the stem. *Listera* → page 173.
        2. Leaves growing from the base of the stem. *Liparis* → page 182.
    C. Plants typically with several leaves at flowering time, arranged alternately on an upright stem. Lip with a saccate structure near the base. Flowers in a slender cluster above the leaves.
        1. Tip of lip long and broad, conspicuously notched. Opening to saccate spur obscure. *Coeloglossum viride* → page 116.
        2. Tip of lip triangular. Base of lip saccate, opening of saccate area conspicuous. *Epipactis helleborine* → page 113.
    D. Plant typically with few to several small leaves at flowering time. Flowers few, each growing singly from an axil. *Triphora trianthophora* → page 105.

Triphora

Triphora trianthophora (Swartz) Rydberg

Etymology: *Triphora: tri* = "three," *phora* = "bearing"; *trianthophora: tri* = "three," *antho* = "flower," *phora* = "bearing."

Common names: three-birds orchid, nodding pogonia.

Description: Leaves 2–8, 6–15 mm (¼–⅗") long, broadly egg-shaped, the base clasping and partially enclosing the stem, alternate, green often with purplish tinges. **Stem** 7.5–20 or up to 28 cm (3–8 or up to 11") tall, color variable from olive to purplish, with 1–6 (usually 3) flowers. **Flowers** about 20 mm (⅘") tall, 25 mm (1") wide, and 15 mm (⅗") long, white often tinged with pink, each flower rising from a separate upper leaf axil. **Sepals** about 14 mm (½") long, lance-shaped, dorsal sepal erect, lateral sepals spreading and often slightly down-curved at the tips, white or occasionally pinkish; **petals** just slightly shorter than sepals, oblong to lance-shaped, curved forward above the lip, white often with pinkish tinges; **lip** about 13 mm (½") long, three-lobed, all 3 lobes somewhat oval, middle lobe largest with a wavy margin and 3 rough greenish ridges down the center, white often with pinkish tinges.

Habitat: in the northeast *Triphora* inhabits predominantly beech woodlands. It grows in rich humus, often in somewhat moist locations such as depressions, the downhill base of boulders, and near streams.

Flowering season: fourth week of July to first week of September (second to fourth week of August).

Range: from Florida to eastern Texas north up the Mississippi River drainage and the eastern states to the southern Great Lakes and southern Maine. In the northeast *Triphora* is found throughout the area except for northernmost New York, New Hampshire, Vermont, and all but the south tip of Maine. *Triphora* is usually considered to be quite rare and localized, but its small size and inconspicuous appearance make it likely that additional (as of yet undetected) populations also occur.

Comments: Of the approximately one dozen recognized *Triphora* species, five occur in the United States. Of these, four confine their range to the warmer weather of Florida. Only one species is known to have spread to the cooler northern climes, and that hardy species may be found over most of the eastern third of our country.

Both the genus and species name of this orchid refer to the number "three," most likely a reference to the number of blooms found on a typical plant. The common name of "three birds" suggests white birds of beauty representing the three stages of life. At the summit we find the bud-egg nestled and waiting to "hatch." Springing from a lower branch but reaching upward to fly above the rest of the plant (nest) is a bird-orchid in prime physical condition, while below roosts an older individual, whose drooping posture indicates that, having achieved procreation, she is ready to fold her wings to make way for the next generation. The alternate common name of nodding pogonia refers to both the old genus *Pogonia* (to which this species was assigned in the 1800s) and to the manner in which the flowers nod or droop once pollination has been effected.

Finding *Triphora* in bloom can be one of nature's most difficult quests because of several unique factors in their natural history. I sometimes say I was extremely lucky to locate *Triphora* in bloom on my first attempt, but in reality my luck was the product of good advice and some solid directions supplied by Paul Martin Brown. Like all our orchids, *Triphora* can only begin and sustain life by sharing a subterranean symbiotic relationship with a fungus. Three birds continues this relationship into the mature life stages, where it lives symbiotically in moist beech humus. This lifestyle may be the safest and surest way for *Triphora* to obtain needed nutrients, but much to the frustration of wildflower lovers, it allows this temperamental orchid to survive out of sight for long periods of time and to restrict its flowering efforts to only those years when a specific set of environmental conditions occur. All the factors that trigger *Triphora* flowering are not understood, but Luer (1975) mentioned temperature as an important influence.

Like most orchids, *Triphora* blooms from bottom to top. Each bud is located in an axil. The lowest bud on the stalk flowers first. Before beginning to bloom the bud twists upward into an erect position, perhaps to display an invitation to potential airborne pollinators. Although the flowers do not open until between noon and three, most blooms have begun to droop by dusk. By the following day the first flower is already hanging downward and fading, while its previously upright position has been taken over by the next bud/flower in line going up the stem. I have heard from some lucky individuals who have witnessed *Triphora* with all three birds in bloom at the same time, but despite several attempts I have not observed this miniature floral flocking.

Another interesting aspect of *Triphora* flowering is that it takes place in "waves." It appears that in excess of 90 percent of mature buds in a colony or hillside bloom simultaneously, with a hiatus of from usually one to two weeks between waves, just one more factor making it uniquely difficult for the average naturalist to observe three birds "in flight."

Galearis

Galearis spectabilis (Linnaeus) Rafinesque

Etymology: *Galearis* = "helmet," in reference to the shape of the combined sepal/petal hood that shelters the lip; *spectabilis* = "showy or grand," as in *spectacular*.

Common names: showy orchis, purple-hooded orchis.

Description: Leaves 2, 7.5–20 cm (3–nearly 8″) long and about half as wide, oblong to egg-shaped, basal, appearing opposite, sheathing the stem, dark green. **Stem** 10–35 cm (4–nearly 14″) tall, smooth, thick, 4–5 angled in cross-section, with usually a pair of leaf-like sheaths below the leaves, green. **Inflorescence** ⅓–½ or more of the stem, usually 5–10 cm (2–4″) long, with 3–8 or up to 15 loosely arranged flowers. **Flowers** about 20–25 mm (⅘–1″) tall, 12 mm (⅖″) wide and long, white and pale purplish. **Sepals** 13–20 mm (½–⅘″) long, narrowly lance-shaped; **petals** about 15 mm (⅗″) long, long and slender, sepals and petals pale purplish, sepals converging with the petals to form a sheltering hood over the lip; **lip** 13–20 mm (½–⅘″) long, egg-shaped with a somewhat wavy margin, pendent, white; **spur** stout, slightly thicker near the tip, about as long as the lip.

Habitat: This orchid grows from rich humus primarily in beech and maple forests. It is commonly associated with ravines and limestone soils.

Flowering season: third week of May to third week of June (first to second week of June).

Range: The showy orchis is found throughout most of the northeastern quarter of the United States south through the mid-Atlantic states. In the northeast it is absent only from the northern third of Maine.

Comments: Earlier in this century both *Galearis spectabilis* and *Amerorchis rotundifolia* were believed to be American representatives of the European genus *Orchis*. In more recent times taxonomists have established *Amerorchis* as a monotypic (one member) genus and included two species in *Galearis*.

Amerorchis

Amerorchis rotundifolia (Banks ex Pursh) Hultén

Etymology: *Amerorchis: Amer* = "American," *orchis* = "testicle," referring to the pair of tuberlike growths found beneath some closely related orchids (but not present in this species). In classical mythology Orchis was the son of a minor deity who was murdered for misbehavior while attending a festival of Bacchus. Considering Bacchus' lurid reputation, we can only guess as to what it took to offend him to that extent. Tradition states that the first orchis sprang from soil soaked in the slain reveler's blood; *rotundifolia: rotund* = "round," *folia* = "leaf," referring to the single nearly circular leaf of this species.

Common names: small round-leaved orchis, little round-leaf, one-leaf orchis.

Description: Leaf solitary, 5–7.5 or up to 10 cm (2–3, up to 4″) long, nearly circular to oval, basal to subbasal, sheathing the stem, dull green. **Stem** 15–25 or up to 35 cm (6–10 or up to nearly 14″) tall, smooth, with usually a pair of leaf-like sheaths below the leaf, green to yellow-green. **Inflorescence** ¼–⅓ of the stem, 4–8 cm (1⅗–3″) long, with 5–10 or up to 15 flowers, flowers loosely arranged and somewhat one-sided on the stem. **Flowers** about 12–14 mm (⅖–½″) tall, 15–17 mm (⅗–⅔″) wide, and 13 mm (½″ inch) long, broad and showy, white to pinkish white with pinkish purple spots. **Sepals** 8–10 mm (³⁄₁₀–⅖″) long, dorsal sepal egg-shaped, lateral sepals lance-shaped, white to pinkish white; **petals** 5–6 mm (⅕–¼″) long, lance-shaped, white to pinkish white sometimes with a few purplish markings; dorsal sepal and petals converging to form a hood over the lip, lateral sepals spreading; **lip** about 8–10 mm (³⁄₁₀–⅖″) long, 3-lobed, middle lobe the largest, wide and indented in the center, white to pinkish white with pinkish purple spots; **spur** slender and about 5–6 mm (⅕–¼″) long.

Habitat: cold, moist cedar woods to mossy spruce and tamarack bogs, often associated with limestone soils.

Flowering season: second week of June to first week of July, occasionally later (probably extirpated in New York, historically the third and fourth week of June).

Range: This northern orchis is found across most of Canada and Alaska, south along the Rocky Mountains to northwestern Wyoming and south to northern New York, Vermont, New Hampshire, and Maine in the east. It is also found in Newfoundland. The northeast was originally a southern limit for this species, but it has apparently been extirpated from all but northern Maine in this region.

Comments: Although it has escaped this pretty little orchid's notice, mankind has held several opinions about the proper taxonomy of this species. Originally listed as an *Orchis,* it was soon moved to the *Habenaria/Platanthera* genus by virtue of its multiflowered stem and slender lip spur. Next it was shifted back to *Orchis,* to demonstrate a suspected "little sister" kinship to the somewhat similar *Orchis* (now *Galearis*) *spectabilis*. Current theory holds that *A. rotundifolia* is sufficiently unique to warrant placement in its own monotypic (one member) genus. This represents quite a reversal in family fortunes, having been moved from the once very large *Habenaria* group to now standing alone as the sole representative of its kind.

Epipactis

Epipactis helleborine (Linnaeus) Crantz

Etymology: *Epipactis,* from an ancient Greek reference (about 350 B.C.) to a plant which was credited with the ability to curdle milk, which may or may not have been from this genus; *helleborine* = "like a hellebore."

Common names: hellebore, helleborine orchid, broad-leaved orchid.

Description: Leaves 3–10, 4–15 cm (1³⁄₅–6″) long and about half as wide, egg-shaped to broadly lance-shaped with bases sheathing the stem, strongly veined, spaced evenly and alternately on the stem, becoming somewhat smaller higher on the stem, green. **Stem** usually 25–60 or up to 100 cm (10–24 or up to 39″) tall, somewhat pubescent especially on the upper portions, green. **Inflorescence** ⅓–½ of the stem, usually 10–35 cm (4–14″) tall, with 15–35 or up to 50 flowers,

flowers loosely arranged and somewhat one-sided on the stem. **Flowers** about 15 mm ($3/5''$) tall and 15–18 mm ($3/5$–$7/10''$) wide, pale green to olive green with purplish and reddish-brown tints, growing outward from the stem but often soon drooping. **Sepals** and **petals** similar, about 10 mm ($2/5''$) long, egg-shaped to lance-shaped, all parts spreading but frequently oriented somewhat forward at the tips, pale green to olive green, sometimes with purplish tints and edgings, petals often with purplish veining; **lip** about 10 mm ($2/5''$) long, constricted in center, basal portion bowl-like, forward portion triangular and downcurved, color similar to but usually lighter than petals, inside of "bowl" brown to purplish-brown.

Habitat: highly variable, found in all types of woods and around trees in most types of soil.

Flowering season: third week of July to third week of August (third week of July to third week of August).

Range: The helleborine is an escaped orchid of Eurasian origin. At this time its range covers most of the northeast, save much of Maine. It has also spread north to southern Ontario, west to Minnesota and south through Pennsylvania. It is the opinion of many that this range will continue to expand in the coming decades. In its home hemisphere it is found through most of south and central Europe eastward to India and Mongolia. It is also found in Africa; in Algeria, Tunisia, and Morocco.

Comments: The helleborine orchid was first reported on our continent from a station near Syracuse, New York, in 1879. We often tend to think of alien introductions only in terms of the "weeds" that inhabit our gardens and lawns. That erroneous point of view may or may not have been responsible for an assertion by the first American observer of this orchid, Mrs. M. O. Rust, that she believed it to be indigenous to the Syracuse area. How it arrived will probably never be known, but since that arrival *Epipactis* has become the commonest woodland orchid throughout much of its still-expanding range. Because of its nonnative status, this orchid lacks the legal protection afforded native species in most states.

The scientific name of *Epipactis helleborine* raises some interesting issues about this plant. Earlier in this century the name *Epipactis* was used to describe the rattlesnake plantains, which are now known as *Goodyera*. At that time this orchid was known under the older name of *Serapias*. The species name of *helleborine*, "like a hellebore," is also quite interesting, and I have entertained the question about which other hellebore it might be named for. The leaves of the helleborine orchid are strongly veined and pleated in a manner that very closely mimics the leaves of *Veratrum viride*, a poisonous member of the lily family, which is best known under the common name of false hellebore.

Veratrum's name of false hellebore also begs the question of what then was the true hellebore for which it was named? For that we must look to plants such as *Helleborus niger*, an early blooming member of the *Ranunculaceae* which we plant in our gardens under the common name of the Christmas rose. I see little physical resemblance between the two, yet it is possible that the name may have been created in reference to chemical qualities the two species share, for both are dangerously toxic. Whatever the true origins for a naming process which has now become obscured by time, conventional wisdom and tradition hold that both *Epipactis* and *Veratrum* owe their namesake to an old world species of *Helleborus*.

Coeloglossum

Coeloglossum viride (Linnaeus) Hartman var. **virescens** (Muhlenberg) Luer

Etymology: *Coeloglossum: coelo* = "hollow," *glossum* = "tongue;" hollow tongue probably referring to the sack-like base of the lip; *viride* = "green," as in virid or verdant growth; *virescens* = "becoming green."

Common names: long-bracted orchid, frog orchid, satyr orchid.

Description: Leaves 2–5, 5–15 cm (2–6″) long, lance-shaped, arranged alternately on the stem with the largest leaves in the lowest positions, green. **Stem** 20–60 up to 80 cm (8–24 up to 31″) or more tall, smooth, green. **Inflorescence** about ⅓ of the stem length, usually 7.5–18 cm (3–7″) long, with 5–25 or more flowers. **Flowers** about 14 mm (½″) tall and 6 mm (¼″) wide, greenish, each flower growing from the base of a floral bract which on the lower portion of the inflorescence greatly exceeds the combined length of the flower and its ovary/stalk. Dorsal **sepal** about 5 mm (⅕") long, egg-shaped, green, lateral sepals about 5 mm (⅕") long, wing-shaped, green; **petals** about 4 mm (⅙") long, very slender, pale green; sepals and petals growing forward to form a sheltering hood over the lip; **lip** about 6–10 mm (¼–⅖") long, long and rectangular, pendent to recurved, tip with 2 prominent tooth-like lobes and usually a tiny rounded lobe between them, whitish green often with purplish tints near the base; **spur** short, about 2 mm (1/12") long, sack-like, hidden behind the lip.

Habitat: usually moist to wet woodlands, but sometimes also in more open or drier situations.

Flowering season: third week of May to third week of July, a few into early August (first and second week of June).

Range: *Coeloglossum viride* as a species is found over most of Europe, much of Asia, most of Canada, and in mountainous or cool regions of the United States, south to New Mexico in the west and North Carolina in the east. In this hemisphere, *var. virescens* is found from Newfoundland to the Aleutian Islands southward. It is found throughout the northeast.

Comments: The long floral bracts make this species superficially similar to *Platanthera flava*, but *C. viride*'s deeply notched lip is easily contrasted with *P. flava*'s tongue-shaped lip. Until recently both species were listed as members of the old genus *Habenaria*, which has now been split into several separate genera. *Coeloglossum* is currently viewed as a single species genus, while *Platanthera* contains a great number of orchids previously recognized as *Habenarias*.

Platanthera

Etymology: *Platanthera: plat* = "wide," *anthera* = "anther," together referring to the broad anther affixed to the top of the column, which is an important classification characteristic.

The genus *Platanthera* is one of four genera that were previously lumped together as the *Habenaria* orchids. Because a majority of older orchid books refer to this genus name, it is important to understand something of it. The name *Habenaria* refers to a "strap" or "rein," a term originally applied to the shape of the spurs and later to the lips of various orchids in this group. The popular name of "rein orchids" was widely used earlier in this century, and the common names of many species still retain this term today.

The original catch-all genus of *Habenaria* was made up of about 500 species, most of them tropical in distribution. After the genus was split the only true *Habenaria*s remaining in the United States were several species residing in Florida and the southeastern coastal plains. Another genus emerging from this split was *Piperia,* a small group of green-flowering orchids found mainly in the northwest, although one species can be found as close as Ontario. *Coeloglossum viride,* the single representative of its genus, is found throughout the area covered by this book. About fifteen or so other species of orchids (and their associated hybrids) that are found in the northeast and were once classified as *Habenaria*s have been transferred into *Platanthera*. As it is defined today, the *Platanthera*s are a group of about 200 species of orchids found mainly in temperate, or colder, climates. The typical Platanthera is a summer (June, July, and August) blooming orchid which supports multiple small to medium-sized flowers at the upper extent of a single stem. The lip may be simple, small and green, or wide, showy and heavily fringed. At the base of the lip is a well-developed spur. Having a lip spur is a characteristic shared in the northeast by only four other genera; *Tipularia, Galearis, Amerorchis,* and *Coeloglossum.* The hollow spur is the repository of sweet nectar which entices insects into the heart of the flower, where they inadvertently (by their standards) take part in the pollination process.

The numerous *Platantheras* can be easily broken down into four smaller subgroups, for the convenience of our own logically minded species. First and most famous are the fringed orchids, seven species known for their elaborately designed lips. The three round-leaved orchids are unusual in that they are better known for their unique plate-like foliage than for their flowers. The *dilatata-hyperborea-huronensis* complex is made up of common green and white orchids familiar to bog-walkers across the northern United States and Canada. The remaining three species are the green orchids, often overlooked jade or emerald orchid jewels, which are so small we need a jeweler's glass to explore their floral beauty.

The *Platantheras* are surpassed only by the lady's slippers in terms of popularity. Many of its members are common, so they are fondly remembered as symbols of success from our earliest orchid hunting days. Many are so beautiful in form that they cannot help but be permanently impressed on our memories. They are at once familiar, fascinating, and add variety to our outdoor experience. It is no wonder so many nature lovers refer to them as our "floral friends."

There is one quick note we must make before entering into the species descriptions. This genus is known for its spurs. Spurs vary greatly in length and shape, with some being straight and others curved. To sidestep the confusion that would otherwise occur, they have not been taken into account or included in the overall flower dimensions. "Tall" for the Platantheras will be the distance from the tip of the dorsal sepal to that part of the lip most opposite it, regardless of the flower's orientation on the stem. Some orchids, such as *P. dilatata,* have a lip whose tip is so curled or hooked upwards that the height of the entire flower could conceivably be less than the length of the lip if it was laid out flat!

Key to the Platanthera Orchids
(continued from page 104)

I. Lip heavily fringed.

II. Lip not fringed. → page 140.

I. Lip 3-lobed.
   A. Flowers purple, 12–22 mm (²/₅–⁴/₅″) tall and wide.
   1. *P. psycodes*
   2. *P. grandiflora* → page 122.

   B. Flowers creamy white to greenish white.
   3. Flowers creamy white. Fringe cuts about halfway into lobes. Petals broadly paddle-shaped. *P leucophea* → page 127.
   4. Flowers greenish white. Fringe cuts almost to base of lobes. Petals narrow. *P. lacera* → page 129.

II. Lip single-lobed
   A. Flowers white.
   5. *P. blephariglottis* → page 131.

   B. Flowers yellow.
   6. Flowers average 18 mm tall × 8 mm wide. Spur 20–30 or more mm (⁴/₅–1¹/₅″ or more) long. *P. ciliaris* → page 133.

   7. Flowers average 10 mm tall × 7 mm wide. Spur 8 (³/₁₀″) long. *P. cristata* → page 135.

The Purple Fringed Orchids

Two species of purple fringed orchids are found in the area covered by this book, and how to differentiate the two has been a matter of considerable debate. If I never left the Adirondack Mountains I could be justified in feeling secure in my ability to differentiate this perplexing duo. *Grandiflora* would be larger, more loosely flowered, a deeper purple and more shade tolerant than the smaller, paler, densely flowering, acid tolerant, and later blooming *psycodes*. Some friends whose powers of observation I respect too well to doubt will argue the validity of any or all of these characteristics. The purple fringed orchids found in the northeast, *Platanthera psycodes* and *P. grandiflora,* are commonly referred to as the small purple fringed orchid and the large purple fringed orchid. In the case of these two very similar species, these common names may be a little misleading to some people. Both plants are of similar height, and in many areas it is actually the small purple fringed orchid that is taller than its large-named counterpart. The names large and small actually refer to the size of the flower, so perhaps using large-flowered and small-flowered purple fringed orchids as names would have been a bit more accurate. Although measuring flower sizes will frequently separate the two species, there is enough of an overlap that this method cannot be relied upon for serious identification efforts. For an excellent indepth examination of this interesting species pair, I would refer the reader to Warren P. Stoutamire (1974). Information from this article regarding the pollinaria of these two species has been incorporated into the following key.

Key to the Purple Fringed Platantheras
(continued from page 120)

I.
- Flower usually 12–15 mm (²/₅–³/₅″) tall and wide.
- Pollinaria 1.5–2 mm (about ¹/₁₂″) long, separated at the bases by 1–1.5 mm (¹/₂₅–¹/₁₂″).
- Opening to spur elongated laterally and slightly constricted near the center.
  P. psycodes → page 123.

II.
- Flower usually 20–22 mm (⁴/₅″) tall and wide.
- Pollinaria about 3 mm (⅛″) long, separated at the base by about 4–5 mm (⅙–⅕″).
- Opening to spur circular, unobstructed.
  P. grandifolia → page 125.

Platanthera psycodes (Linnaeus) Lindley

Etymology: *psycodes* = "butterfly-like," probably an allusion to the broad wing-like lobes of the lip.

Common names: small purple fringed orchid, lesser purple fringed orchid, butterfly orchid.

Description: Leaves 2–5, 5–20 cm (2–nearly 8″) long, lance-shaped to somewhat oblong, alternate, green. **Stem** 15–70 or more cm (6–27″ or more) tall, smooth, green. **Inflorescence** about ³⁄₁₀ of the total stem, usually 5–20 cm (2–8″) long and 3–4 cm (1⅕–1⅗″) wide, with 25–50 or more usually densely arranged flowers. **Flowers** 12–15 mm (⅖–⅗″) tall and wide, usually pale purple to lilac but commonly deep purple or white, fragrant. **Sepals** and **petals** similar, 5–7 mm (⅕–¼″) long, somewhat egg-shaped, margins of sepals smooth, margins of petals minutely toothed, dorsal sepal and petals erect or growing slightly forward, lateral sepals recurved; **lip** 7–12 mm (¼–⅖″) long and slightly wider, composed of three broadly fan-shaped lobes, lobes heavily fringed, lateral lobes usually flat or recurved; **spur** 12–20 or more mm ($2/5$–$4/5$ or more″) long, slender and often slightly thickened at the tip, the opening to the base of the spur is often elongated laterally and slightly constricted near the center.

Habitat: usually found in moist to wet situations from meadows and sphagnum fens to the edges of creeks and rivers.

Flowering season: third week of July to fourth week of August, a few into September (fourth week of July to third week of August).

Range: from Newfoundland to New Jersey, west to Ontario and Tennessee. It is found throughout the northeast.

Comments: The small purple fringed orchid is commoner in most locations than its larger look-alike relation. It often occurs in large colonies, and within each colony one is likely to encounter flowers of every shade of color from white to purple.

Platanthera grandiflora (Bigelow) Lindley

Etymology: *grandiflora: grandi* = "large," *flora* = "flowered," this species has larger flowers than *P. psycodes*.

Common names: large purple fringed orchid, greater purple fringed orchid, large butterfly orchid, early purple fringed orchid.

Description: Leaves 2–6, 5–20 or more cm (2–8″ or more) long, lance-shaped to somewhat oblong, alternate, green. **Stem** 15–65 or more cm (6–25″ or more) tall, smooth, green. **Inflorescence** about ³⁄₁₀ of the total stem, 7.5–25 cm (3– nearly 10″) or more tall, and 4–6 cm (1³⁄₅–2²⁄₅″) wide, with usually 15–25 loosely arranged flowers. **Flowers** about 20–22 mm (⁴⁄₅″) tall and almost as wide, usually purplish but often pale purple to white, somewhat fragrant. Dorsal **sepal** 6–9 mm (¼–⅓″) long, oblong to oval with smooth margins; lateral sepals 6–10 mm (¼–²⁄₅″) long, broadly lance-shaped with smooth margins; **petals** 6–10 mm (¼–²⁄₅″) long, oblong to nearly rectangular with minutely toothed margins; dorsal sepal and petals erect or growing slightly forward, lateral sepals recurved; **lip** about 10–19 mm ($^2/_5$–$^3/_4$″) long and slightly wider, composed of three broadly fan-shaped lobes, lobes heavily fringed, lateral lobes usually curve slightly forward; **spur** 20–30 mm ($^4/_5$–1$^1/_5$″) long, slender, the opening at the base of the spur is usually circular or sometimes slightly elongated vertically.

Habitat: usually found in moist to wet situations from open woodlands to swamps.

Flowering season: fourth week of June to first week of August (second to the fourth week of July).

Range: from Newfoundland to New Jersey, westward to the central Great Lakes area and Tennessee. It is found throughout the northeast except for Long Island.

Comments: The large purple fringed orchid is frequently found along roadsides in wooded areas. Unfortunately, this leaves them vulnerable to damage from road enlargements and "improvements," casual flower pickers who are not aware of the protected status of this species, and summer roadside mowing, which destroys the reproductive abilities by cutting down the stems long before viable seed has formed.

Platanthera leucophaea (Nuttall) Lindley

Etymology: *leucophaea* = "whitish," referring to the creamy white flowers.

Common names: prairie white fringed orchid, prairie fringed orchid.

Description: Leaves 2–5, 10–20 cm (4–nearly 8″) long, narrowly lance-shaped, alternate, green. **Stem** 45–75 or up to 120 cm (18–30 or up to 47″) tall, smooth, green. **Inflorescence** about ⅕ of the total stem, usually 7.5–13 or up to 20 cm (3–5, up to nearly 8″) long, with 10–20 or up to 25 loosely arranged flowers. **Flowers** about 28 mm (1¹⁄₁₀″) tall and 21 mm (⅘″) wide, creamy white, and fragrant, especially at and after sunset. **Sepals** similar, 7–16 mm (¼–⅗″) long, egg-shaped, white with a pale greenish tinge; **petals** 7–16 mm (¼–⅗″) long, broadly paddle-shaped but notched at the tip, margins minutely toothed, creamy white; sepals and petals growing somewhat forward around the base of the lip; **lip**

15–30 mm (³/₅–1¹/₅″) or more long and wide, composed of three broadly fan-shaped lobes, lobes heavily fringed, the fringe cutting more than half way to the base, especially on the lateral lobes, pendent, creamy white; **spur** 30–50 mm (1¹/₅–2″) or more long, slender and somewhat thickened near the tip.

Habitat: open moist areas such as lakeshores, fens, wet meadows and, as the name suggests, prairies.

Flowering season: second week of July to second week of August (possibly extirpated, historically second to third week of July).

Range: from western New York to Iowa, Missouri, and Oklahoma. There is a disjunct population in northern central Maine. Since the western New York populations are possibly extirpated, Maine may represent the only surviving population in the northeast, making this species one of the scarcest orchids in this area.

Comments: One turn of the century orchid book refers to the prairie fringed orchid as a glorified form of the ragged fringed orchid. To me the size and shape of the lip suggest a closer kinship to the purple fringed orchids. If it were possible to carefully color a prairie fringed orchid, most of us would have great difficulty in differentiating it from its royally hued relations.

Platanthera lacera (Michaux) G. Don

Etymology: *lacera* = "torn," as in "lacerate" (to tear) and referring to the ragged, torn appearance of the lip.

Common names: ragged fringed orchid, green fringed orchid.

Description: Leaves 2–5, 7–25 cm (2¾–nearly 10″) long, narrowly lance-shaped, alternate, green. **Stem** usually 30–60 or up to 75 cm (12–24 or up to 30″) tall, smooth, green. **Inflorescence** about ⅓ of the total stem, usually 5–15 or up to 25 cm (2–6 or up to nearly 10″) long, with 20–40 usually loosely arranged flowers. **Flowers** about 20 mm (⅘″) tall and wide, yellowish green to greenish white or dirty white. Dorsal **sepal** about 5 mm (⅕″) long, egg-shaped; lateral sepals about 6 mm (¼″) long, unevenly oblong; **petals** about 6–7 mm (¼″) long, narrow; dòrsal sepal and petals erect or oriented slightly forward, lateral sepals recurved; **lip** 10–16 mm (⅖–⅗″) long and slightly wider, 3-lobed, lobes heavily fringed, lateral lobes fringed nearly to the base, the central lobe is typically fringed only ⅓ of its length, pendent; **spur** 10–20 or more mm (²/₅–⁴/₅″ or more) long, slender.

Habitat: highly variable, from woodlands and moist meadows to fens, swamps, and riverbanks.

Flowering season: third week of June to third week of August, a few into September (fourth week of July to first week of August).

Range: from Nova Scotia to South Carolina, west to Manitoba and Arkansas. It is found throughout the northeast.

Comments: Often dubbed the "least attractive of the fringed orchids," *P. lacera* has been given an undeserved little brother/poor relation reputation. Up close I find these off-white flowers every bit as beautiful as those of its more colorful cousins. If *Aplectrum hyemale* had not already laid claim to the common name of "Adam and Eve," this might have been an appropriate common name for this species, for I often find it in pairs.

Platanthera blephariglottis (Willdenow) Lindley var. **blephariglottis**

Etymology: *blephariglottis: blephari* = "eyebrow or eyelash," *glottis* = "tongue," together referring to the tongue-shaped, heavily fringed lip.

Common name: white fringed orchid.

Description: Leaves 2–3, 5–20 cm (2–nearly 8″) long, narrowly lance-shaped, alternate, green. **Stem** 30–60 cm (12–24″) tall, smooth, green. **Inflorescence** ¼–⅓ of the total stem, usually 4–15 or up to 20 cm (1⅗–6 or up to nearly 8″) long, with 10 to 20 or up to 30 densely to loosely arranged flowers. **Flowers** about 10–15 mm (⅖–⅗″) tall and 10 mm (⅖″) wide, all parts snow white. **Sepals** similar, about 8–9 mm (³⁄₁₀–⅓″) long, somewhat egg-shaped; **petals** about 7 mm (¼″) long, oblong to narrowly rectangular, often minutely toothed at the tip; dorsal sepal and petals erect and oriented slightly forward, lateral sepals fully recurved; **lip** about 10 mm (⅖″) long, nearly oval to tongue-shaped, heavily fringed along the margin, pendent; **spur** 15–23 mm ($3/5$–$9/10''$) long, slender.

Habitat: in the northeast the white fringed orchid is most often found in sphagnum fens, but may also be found in other moist situations.

Flowering season: first week of July to fourth week of August (third week of July to third week of August).

Range: *Platanthera b.* var. *blephariglottis* is found from Newfoundland to New Jersey westward to Michigan. It is found throughout the northeast. Variety *conspicua* is found along all of the southeastern coastal states.

Comments: Variety *conspicua* is the southern counterpart to our northeastern variety of this species. It is much larger, with 25–30 mm (1–$1^{1}/_{5}''$) tall flowers with spurs up to 40 mm ($1^{3}/_{5}''$) long! No matter which variety is encountered, the snow white plumes of flowers never fail to fascinate those lucky enough to encounter them.

Platanthera ciliaris (Linnaeus) Lindley

Etymology: *ciliaris* = "eyelashed," referring to the heavily fringed lip.

Common name: yellow fringed orchid.

Description: Leaves 2–4, 5–20 or up to 30 cm (2–8 or up to 12″) long, lance-shaped, alternate, green. **Stem** 30–75 or up to 100 cm (12–30 or up to 39″) tall, smooth, green. **Inflorescence** $1/5$–$1/3$ of the total stem, usually 7.5–15 or up to 20 cm (3–6 or up to nearly 8″) long, with 25–50 or up to 60 usually densely arranged flowers. **Flowers** about 15–20 mm ($3/5$–$4/5$″) tall and 6–10 mm ($1/4$–$2/5$″) wide, all parts yellowish orange. **Sepals** similar, about 8 mm ($3/10$″) long, somewhat egg-shaped; **petals** about 6 mm ($1/4$″) long, oblong to narrowly rectangular, minutely toothed at the tip; dorsal sepal and petals erect and oriented slightly forward, lateral sepals strongly recurved; **lip** 13–20 mm ($1/2$–$4/5$″) long, oblong, heavily fringed on margin (the fringe accounting for $1/3$–$1/2$ of the total lip length), pendent; **spur** 20–30 or more mm ($4/5$–$1\,1/5$″ or more) long, slender.

Habitat: variable, from open fens and meadows to sandy soils, shaded woodlands, and moist roadsides.

Flowering season: third week of July to fourth week of August, a few into early September (fourth week of July to first week of August on Long Island).

Range: from Vermont to Florida, west to Illinois and Texas. In the northeast it originally ranged throughout except for Maine, northern New York, New Hampshire and Vermont, and eastern Massachusetts. The actual present-day range is spotty at best, with the largest populations surviving in coastal areas. At this time this species is severely threatened in New York State due to human disturbance.

Comments: If we were to bleach the bright coloration from this orchid, we would have difficulty differentiating it from its white fringed relation. Luer (1975) pointed out that distinguishing between the dried herbarium specimens of *P. cilliaris* and *P. blephariaglottis* var. *conspicua* can be very difficult. In earlier times the white fringed was believed to be nothing more than a white from (var. *alba*) of *P. ciliaris*. At least there is no difficulty in separating these species when we meet them in the wild, where their bright fresh colors make identifications instantaneous. Where they coexist in the north, it is claimed that the white fringed begins blooming a few days before the yellow fringed orchid, just the opposite of what is reported from the southeastern states!

Platanthera cristata (Michaux) Lindley

Etymology: *cristata* = "crested," referring to the fringed tip of the petals, which protrude from under the dorsal sepal to display a crest on the upper portion of the flower.

Common names: crested fringed orchid, crested yellow orchid.

Description: Leaves 2–4, 5–20 cm (2–nearly 8″) long, lance-shaped, alternate, green. **Stem** 20–60 or up to 80 cm (8–24 or up to 31″) tall, smooth, green. **Inflorescence** ⅕–¼ of the total stem, usually 5–10 or up to 15 cm (2–4 or up to 6″) long, with 40–80 densely arranged flowers. **Flowers** about 10 mm (⅖″) tall and 6–8 mm (¼–³⁄₁₀″) wide, all parts bright orange. **Sepals** similar, about 4 mm (⅙″) long, nearly round; **petals** about 3 mm (⅛″) long, egg-shaped with fringed margins; dorsal sepal and petals growing forward over the base of the lip, lateral sepals spreading; **lip** 7–8 mm (¼–³⁄₁₀″) long, egg-shaped, heavily fringed on margin (the fringe accounting for ⅓–½ of the total lip length), many fringe filaments forked, pendent or curved slightly upward; **spur** about 8 mm ($^3/_{10}''$) long, slender, opening at the base of the spur like an inverted triangle.

Habitat: usually found in sandy soils in meadows or open pine woods. Can be found growing quite close to the ocean in some locations.

Flowering season: fourth week of July to second week of August (fourth week of July to first week of August on Long Island).

Range: This orchid is found on the coastal plain from Massachusetts around Florida to Texas, and inland to Arkansas and Tennessee. In the northeast its range is restricted to eastern Massachusetts, Rhode Island, Connecticut, and southeastern New York.

Comments: *Platanthera cristata* is considered rare and endangered in New York State. On Long Island it is a bright orange, definitely darker and richer than the nearby *P. ciliaris*, the yellow fringed orchid. Further south, in Virginia, I have seen entire populations of *P. cristata* whose flowers were a paler yellow than New York's *P. ciliaris*.

Platanthera cristata, yellow form, Virginia

Platanthera pallida P. M. Brown

Etymology: *pallida* = "pale," and refers to the color of the flowers when compared to typical *P. cristata*.

Common names: pale fringed orchid, pale cristata.

Description: Leaves 2–3, up to 25 or 30 cm (10–12″) long, lance-shaped, alternate, green. **Stem** 29–65 or up to 84 cm (11½–25½ or up to 33″) tall, smooth, green. **Inflorescence** about ¼–⅓ of the total stem, 10–20 or up to 27 cm (4–8 or up to 10½″) long, with 24–80 or up to over 100 densely arranged flowers. **Flowers** about 4–5 mm (⅙–⅕″) tall and 3 mm (⅛″ inch) wide, all parts creamy white. **Sepals** similar, about 3 mm (⅛″) long and wide, nearly round; **petals** about 3.25 mm (⅐″) long, egg-shaped with fringed margins; dorsal sepal and petals growing forward over the base of the lip, lateral sepals recurved; **lip** about 5–6 mm (⅕–¼″) long, strap-shaped, heavily fringed on the margin (the fringe accounting for about ½ of the total lip length), many fringe filaments forked, lip usually recurved or occasionally pendent; **spur** 5–6 mm (⅕–¼″) long, slender, opening at the base of the spur T-shaped to keyhole-like.

Habitat: dry, interdunal hollows quite close to the ocean, under sheltering *Pinus rigida* and associated trees and shrubs.

Flowering season: fourth week of July to second week of August.

Range: limited to a small area on eastern Long Island. If *P. pallida* is recognized as a distinct species, it will have the smallest range of any American orchid.

Comments: Tucked safely within selected sheltered interdunal hollows of eastern Long Island, a small population of cream-colored orchids survives the rigors of the harsh beach environment. Paul Martin Brown (1992) has proposed that these plants be awarded species status. The major characteristics claimed for this orchid are the pale coloration, the back growing/arching lateral sepals and lip, and the highly specific habitat. At this time the debate has not yet been fully resolved about whether or not these orchids represent a possible hybrid, a distinct color form, a variety in the process of evolving into a different species, or a distinct species. Irrespective of our opinions concerning its possible origins and relations, this interesting orchid continues along on its unique evolutionary path. The above description takes no sides in this debate but is intended to describe the orchid as it was presented in Brown's 1992 article.

Platanthera × andrewsii

Hybridization among the Fringed Platantheras

Whenever closely related species of plants coexist in the same habitat, hybridization is a possibility. The following hybrids have been observed among the fringed orchids:

1. *P. × andrewsii*, from *P. lacera* with *P. psycodes*. *andrewsii* refers to Albert LeRoy Andrews (1878–1961), who first publicized this hybrid. One of the commoner hybrids in the northeast, specimens typically combine the deeply "torn" lip of *P. lacera* with some degree of the purplish coloration of *P. psycodes*. Of special interest is a pure white specimen photographed in the Adirondack Mountains by Evelyn Greene. The white coloration may have resulted from either one parent being a white *P. psycodes* (such plants exist only a few miles away) or the bloom may have been exhibiting albinism. Recorded flowering dates: July 20 in Lewis County, New York, and July 24 in Warren County, New York.
2. *P. × bicolor*, from *P. blephariglottis* with *P. ciliaris*. *bicolor*: bi = "two," color = "color," referring to the two colors which mix to create this hybrid. We have already seen that

except for the characteristic of color the white fringed and the yellow fringed orchids are physically nearly identical. Hybrids between these two species can be identified by their creamy color, too dark to be *P. blephariglottis* and too pale to be *P. ciliaris*. Recorded flowering dates: August 15 in Suffolk County, New York.

3. *P.* × *canbyi*, from *P. blephariglottis* with *P. cristata*. *canbyi* refers to William Canby, who first publicized this hybrid. Physically, this hybrid looks like a slightly large, pale *P. cristata* with an unusually long spur (for *P. cristata*). Correll (1950) and others have suggested that this hybrid may be a case of "mother's baby, papa's maybe." *P. canbyi* is found growing with *P. cristata,* and there is at least a question as to whether the absent parent is *P. blephariglottis* or *P. ciliaris*. I have been shown *P. canbyi* on Long Island. These plants were mixed in with *P. cristata* and just down the road from *P. blephariglottis,* but to be fair they were also not all that far from a stand of *P. ciliaris*. If it is ever proven that these crosses do involve *P. ciliaris,* then *P. canbyi* will become synonymous with *P.* × *channellii*. Recorded flowering dates: August 3 on Long Island.

4. *P.* × *channellii*, from *P. ciliaris* with *P. cristata*. *channellii* in honor of Robert B. Channell, who drew attention to this hybrid. Hybrids between these two species display color, floral size, and spur lengths that are intermediate between that of the parents. Recorded flowering dates: I am not aware of any records in the northeast, but this cross is hypothetically possible in Connecticut and on Long Island. Late July through early August would be the likely blooming season.

Key to the "Nonfringed" Platanthera
(continued from page 120)

Flowers green, yellowish green, greenish white, or white.

I. Leaves 2, basal, nearly round → page 141.

II. Leaves several, arranged alternately on an upright stem, much longer than wide.

A. Lip long and tapering to the tip. Lateral sepals narrow, spreading outwards, giving the flower a "winged" appearance → page 150.
   B. Lip long and bluntly tongue-shaped at the tip, with a "wartlike" growth near the base. Lateral sepals egg-shaped, recurved → page 157.
III. Leaf usually single.
   A. Lip nearly rectangular but slightly wider and shallowly 3-lobed at the tip → page 159.
   B. Lip narrow and tapering to the tip → page 161.

The Round-Leaved Orchids

Three orchids, *P. hookeri*, *P. orbiculata*, and *P. macrophylla*, make up the round-leaved orchid group. They are members of that unique group of orchids whose foliage is far more eye catching than the flowers. For these three orchids each plant seems to set a dinner table for a pair of woodland spirits because in each species the most conspicuous physical feature is the pair of plate-like leaves lying on the forest floor. The stem rises between the "plates" to complete the illusion, a green candle to provide light for the spirits' romantic nocturnal feasts!

Key to the Round-leaved Platanthera*

   (continued from page 140)

I. Spur length less than 28 mm (1$\frac{1}{10}$″)
   A. Lip triangular, tip curved up like a hook. Flowers yellowish green.
      P. hookeri → page 142.
   B. Lip long and narrow, pendent. Flowers whitish.
      P. orbiculata → page 146.
II. Spur length 28 mm (1$\frac{1}{10}$″) or longer.
   A. Lip long and narrow, pendent. Flowers whitish.
      P. macrophylla → page 147.

* This key incorporates information from *The Species Pair Platanthera orbiculata and P. macrophylla*, by Reddoch and Reddoch 1993.

Platanthera hookeri (Torrey ex Gray) Lindley

Etymology: *hookeri:* in honor of William Jackson Hooker, an English botanist who lived from 1785–1865.

Common names: Hooker's orchid, Hooker's round-leaved rein-orchid, Hooker's rein-orchid.

Description: Leaves 2, 7–15 cm (2¾–6″) long and almost as wide, nearly round, lying nearly on the ground or raised slightly higher, growing from the base of the stalk, green. **Stem** 20–40 cm (8–nearly 16″) tall, smooth, vertically ridged, usually without bracts, green. **Inflorescence** ⅖–½ of the total stem, 10–20 cm (4–nearly 8″) long, with 10–25 loosely arranged flowers. **Flowers** 10 mm (⅖″) tall and 5–8 mm (⅕–³⁄₁₀″) wide, all parts yellowish green. Dorsal **sepal** 8–10 mm (³⁄₁₀–⅖″) long, tear-shaped, growing forward over the base of the lip; lateral sepals 9–12 mm (⅓–⅖″) long, narrowly lance-shaped, completely recurved; **petals** 7–9 mm (¼–⅓″) long, narrow, somewhat sickle-shaped but turned outward at the tips, growing forward and crossing under the dorsal sepal; **lip** 9–12 mm (⅓–⅖″) long, triangular, growing downward and forward at the base but curving up like a hook at the tip; **spur** 15–25 mm (⅗–1″) long, slender.

Habitat: typically dry mixed or coniferous forests, occasionally found in more moist situations.

Flowering season: fourth week of May to first week of July, (first to second week of June).

Range: from Newfoundland south to New Jersey, west to Manitoba and Iowa.

Comments: Hooker's orchid is one of those wildflowers which inspires fanciful associations in the minds of human viewers. Viewed from the side, the prominent curving lip has been compared to everything from an upraised elephant's trunk to a coathook—but remember this hook shape is not the source of the common name! Correll (1950) compares the face of this orchid to that of a gargoyle, with the dorsal sepal curving forward to provide a hood, while the lip curves up to simulate a long and pointed demonic chin.

The Platanthera orbiculata-macrophylla Species Pair

Both of the species listed above were originally described in the early 1800s, *P. orbiculata* by Pursh and *P. macrophylla* by Goldie. Goldie (1822) mistakenly reported that *P. macrophylla* was much larger than *P. orbiculata*. This misconception was to be perpetuated by botanists until very recent times. Early in this century, Ames (1906) still supported the idea that *P. orbiculata* and *P. macrophylla* represented distinct species and prophetically suggested spur length as one differentiating characteristic. Perhaps overwhelmed by the difficulties these two similar species presented, most orchidologists after Ames lumped the two together under the name of *Habenaria orbiculata*. Variety *macrophylla* was recognized as a way to classify those unusually large specimens encountered in New York and around the Great Lakes. Writing in 1993, Allan and Joyce Reddoch clarified both the similarities and the differences between these two orchids and have restored species level recognition to each. The Reddochs' extensive research has formed the basis for the data given here—including the historical information already presented—leaving all orchid lovers in their debt for clarifying and simplifying a previously confusing situation.

One major finding of the Reddochs was how similar the two species really are. Although it is true that when comparing most of the major physical traits *P. macrophylla* is the larger of the two, there is so much overlap between the sizes of "average" plants of both species that measurements such as height and leaf size cannot be counted on for differentiation. What can be counted on, with a reported 99% accuracy, is a measurement of the spur. The Reddochs found that plants with a spur length of less than 28 mm may be classified as *P. orbiculata*, while a spur length of 28 or more mm places the plant within *P. macrophylla*.

Ironically, even though spur length makes the two species easy to separate, their similarities in some ways make the general descriptions more difficult. Rather than provide two descriptions that differ only minimally, this work will employ only one description giving a range of measurements that, unless otherwise noted, fall within the now recognized size ranges for both these species. Please keep in mind that this work represents the sizes of "average" plants, and that both smaller and larger specimens may be encountered in the field. The reader wishing a more thorough study of size ranges for the component parts of these species is referred to the Reddochs' article.

Platanthera orbiculata (Pursh) Lindley
(photo above)

Platanthera macrophylla (Goldie) P. M. Brown

Etymology: *orbiculata* = "round," as in "orbit," and referring to the leaves; *macrophylla: macro* = "large," *phylla* = "leaf," together referring to the large circular leaves of this species.

Common names: Traditionally, large round-leaved orchid was applied to both these species. The Reddochs suggest retaining this common name for *P. orbiculata* and recommend the name of Goldie's round-leaved orchid for *P. macrophylla*.

Description: Leaves 2, 8–20 cm (3–nearly 8″) long and almost as wide, nearly round and lying flat on the ground, growing from the base of the stem, glossy green above, silvery beneath. **Stem** 25–60 cm (10–24″) tall, smooth, often with a few tiny bracts, green to whitish-green. **Inflorescence** usually $3/10$–$4/10$ of the total stem, 8–20 to 25 cm (3–8, up to nearly 10″)

Platanthera macrophylla

long, with 10–30 or up to 40 loosely to densely [in *P. orbiculata*] arranged flowers. **Flowers** about 15 mm ($3/5''$) wide and 18 mm ($7/10''$) tall (to 25 or more mm [$1''$ or more] tall in *P. macrophylla*), all parts off-white to greenish-white. Dorsal **sepal** 5–6.5 mm ($1/5$–$1/4''$) long and about as wide (up to 9 mm [$1/3''$] in *P. macrophylla*), oriented forward over the base of the lip; lateral sepals 8–11 mm ($3/10$–$2/5''$) long, broadly sickle-shaped, recurved, **petals** 7–12 mm ($1/4$–$2/5''$) long, sickle-shaped, spreading and erect; **lip** 10–15 mm ($2/5$–$3/5''$) long (up to 23 mm [$9/10''$] long in *P. macrophylla*), long and narrow, pendent; **spur** 14–28 mm ($1/2$–$1 1/10''$) in *P. orbiculata,* 28–46 mm ($1 1/10$–$1 4/5''$) in *P. macrophylla,* long and slender, slightly thickened at the tip.

Habitat: *Platanthera orbiculata* is found in habitats ranging from moist mixed forests to wooded swamps. *Platanthera mac-*

rophylla is found mainly on slopes of the mixed maple-beech-hemlock woods typical to much of the northeast.

Flowering season: fourth week of June to first week of August (fourth week of June to third week of July).

Range: *Platanthera orbiculata* is found across Canada from Newfoundland to British Columbia. It is found south to Oregon in the west to Tennessee in the east. *Platanthera macrophylla* has a much smaller range and is found from Newfoundland to Michigan and south to Pennsylvania. Both species are found throughout the northeast, most of the Cape Cod area excepted.

Comments: The two round-leaved orchids discussed here are so similar that over the years many have questioned whether or not they deserve to be separated at the species level. The rationale for recognizing species status lies in differing reproductive organs such as the spurs. Since these differences preclude the likelihood of cross pollination, separate species status for each is inferred. Not long ago I was pondering the huge *P. macrophylla* that grow in the Adirondack Mountains. It was fun to fantasize that some benevolent woodland spirit long ago conferred near immortality on these stately plants, some of which were very young when they were named *Orchis orbiculata* by botanists. Soon we changed their name to *Habenaria* and later *Lysias macrophylla*. Not so long ago, visiting naturalists referred to them as *Habenaria orbiculata* var. *macrophylla*. More recently, visitors have addressed them as *Platanthera macrophylla*. All the while these wise old orchids have stood silently watching the parade of interested but obviously confused humans passing by, while they never once doubted their one true unspoken name.

The Platanthera dilatata-hyperborea-huronensis Complex

A complex, from a botanical point of view, is a group of three or more closely related species of plants. If we were discussing a complex in psychiatric terms, we might be referring to patients or individuals with identity problems. In this case it is not the individual but we outsiders that have the problem with feeling secure in assigning identities. In 1950 Correll referred to *Platanthera hyperborea* as "doubtless the most perplexing species of *Habenaria* within our range." Correll's view is still valid today; most botanists agree that so much variation exists within and between the species of this complex that describing distinct taxa is both difficult and sometimes somewhat arbitrary. *Platanthera dilatata* has been the easiest to identify, with only a few varieties based on lip variations. *Platanthera hyperborea* has been thought to have at least five varieties, and on the west coast the presence of an additional three very similar species compound the problems of the identification process.

When I first became interested in orchids, identifications within this group were easily accomplished, for *P. huronensis* was not at that time recognized as a separate species. If the flowers were pure white and fragrant, the species was *P. dilatata*. If the flowers were green and not fragrant, the species was *P. hyperborea*. So similar in shape are these two species that they were termed the "twin sisters." Any specimens with pale green, somewhat fragrant, flowers was identified as *P. × media*, a presumed hybrid between the two previous species.

Many specimens matching this general description from the Great Lakes area were unusually large and classified as *P. hyperborea* var. *huronensis* (Luer 1975). Many orchidologists now recognize *P. huronensis* as a distinct species, based on its larger size, lighter color and narrower lip. In addition, *P. hyperborea* is at least sometimes self-pollinating, while it is believed *P. huronensis* is probably not (Sheviak, personal communication). This classification scheme explains the large lighter green orchids of the Great Lakes region, but does this mean there are no true hybrids out there, or if there are, how do we differentiate them from *P. huronensis*? As has always been the case with this complex, questions, questions, more and more questions!

Key to the "Wing-sepaled" Platantheras

(continued from page 140)

I. Flowers white, very fragrant, 12–18 mm ($^2/_5$–$^7/_{10}''$) wide.
P. dilitata → page 151.

II. Flowers some shade of green, with moderate or no fragrance.

  A. Flowers about 12 mm ($^2/_5''$) tall and wide, green to yellowish green, not fragrant.
  P. hyperborea → page 153.

  B. Flowers about 14–18 or more mm ($^1/_2$–$^7/_{10}''$ or more) tall and wide, whitish green to pale or bright yellowish green, moderately fragrant.
  P. huronensis → page 155.

Platanthera dilatata (Pursh) Lindley ex Beck var. **dilatata**

Etymology: *dilatata* = "widened," as in "dilated," and refers to the widening at the base of the lip.

Common names: tall white bog orchid, bog candle, fragrant orchid.

Description: Leaves 6–12, 5 to 30 cm (2–nearly 12″) long, narrowly lance-shaped, erect and giving the plant a slender appearance, green. **Stem** 30–60 up to 75 or more cm (12–24 up to 30″ or more) tall, smooth, green. **Inflorescence** ⅓–½ of the total stem, usually 10–30 cm (4–12″) long, with many (up to over 100!) densely arranged flowers. **Flowers** about 12 mm (⅖″) tall if lip is pendent, 12–18 mm (⅖–⁷⁄₁₀″) wide, all parts pure white, very fragrant, fragrance of cloves or the "warm" cooking spices, some say also with a touch of vanilla. Dorsal **sepal** about 5 mm (⅕″) long, egg-shaped; lateral sepals about 6–9 mm (¼–⅓″) long, narrowly lance-shaped; **petals** about 6 mm (¼″) long, sickle-shaped; dorsal sepal and petals grow forward forming a sheltering hood over the lip, lateral sepals spread outward or are slightly recurved; **lip** about 6–9 mm (¼–⅓″) long, long and slender but usually dilated to the point of being nearly circular at the base, the long slender tip curves upward and is often hooked into the "hood" when the flower first opens, later the lip becomes pendent to slightly recurved—the typical flowering stalk exhibits both extremes and all gradations between; the **spur** is slender and about as long as the lip.

Habitat: usually in sunny moist areas such as fens or wet meadows.

Flowering season: third week of June to first week of August, a few through August (third week of June to third week of July).

Range: from Newfoundland and Quebec south to New York and west to Alaska and California. It is found throughout the northeast except for eastern Massachusetts.

Comments: The tall white bog orchid is a truly beautiful orchid that reaches its southern limit on the east coast in New York. In spite of this, it is quite common in many sphagnum fens. As the lip evolves from its upper to lower position, the face of the flower undergoes corresponding changes that have been imagined to represent the white bird of peace, a coat-hook, an elephant (the original white elephant?), and finally an orchid!

Platanthera hyperborea (Linnaeus) Lindley var. **hyperborea**

Etymology: *hyperborea: hyper* = "above or beyond," *borea* = "north." The Hyperboreans were a mythical race of beings from the land of the northern lights (aurora borealis) and the northern wind. As the name implies, this orchid grows north into the arctic.

Common Names: tall green bog orchid, tall northern green orchid.

Description: Leaves 4–6; 5 to 15 or more cm (2–6″ or more) long; narrowly lance-shaped, somewhat erect, green. **Stem** 20–40 or more cm (8–16″ or more) tall, smooth, green. **Inflorescence** about ⅓ of the total stem, usually 6–19 cm (2⅖–7½″) long, with many (up to 75 or more) densely to loosely arranged flowers. **Flowers** about 10–12 mm (⅖″ inch) tall and wide, green to yellowish green, not fragrant. Dorsal **sepal** about 4 mm (⅙″) long, egg-shaped; lateral sepals about 5 mm (⅕″) long, narrowly lance-shaped; **petals** about 5 mm (⅕″)

long, sickle-shaped; dorsal sepal and petals grow forward forming a sheltering hood over the lip, lateral sepals spread outward or are slightly recurved; **lip** up to 6 mm (nearly ¼″) long, lance-shaped to long triangular with a dilated somewhat angular base, the long rounded tip curves upward and is often hooked into the "hood" when the flower first opens, later it becomes pendent to slightly recurved—the typical flowering stalk exhibits both extremes and all gradations between; the **spur** is slender with a thickened and blunt tip.

Habitat: highly variable, from open moist or wet areas to shaded woodlands, occasionally occurring in drier soils. Luer (1975) points out that plants growing in open or sunny situations are more likely to produce a cylindrical, densely flowered inflorescence, while well-shaded plants show a predisposition toward a slimmer loosely flowered inflorescence.

Flowering season: third week of June to second week of August, one of the longest bloom seasons for any northeastern orchid (fourth week of June to second week of August).

Range: *Platanthera hyperborea* is found in Japan, Iceland, Greenland, most of Canada and Alaska, south around the Great Lakes and south to California, Arizona, and New Mexico in the west. It is found throughout the northeast except for southeastern New York, Rhode Island, and eastern Massachusetts.

Comments: The tall green bog orchid is one of our commonest and longest flowering orchids, but it is often overlooked due to its inconspicuous coloration. There must be an aspect of human nature that the brighter colored flowers are more likely to stimulate our imaginations, for green *P. hyperborea* rarely inspires the associations commonly credited to the similarly sized and shaped white *P. dilatata*. Even though there seems to be no "birds" in *P. hyperborea,* many who peer deeply into the flower say they see a little green face looking back!

Platanthera huronensis (Linnaeus) Lindley

Etymology: *huronensis;* referring to Lake Huron and indirectly the Huron Indians, making this the only orchid I know of named for Native Americans.

Common names: tall green bog orchid, tall northern green orchid (These are the same common names also applied to P. *hyperborea.* Presumedly, as more people learn to differentiate the two species, separate common names will also evolve.)

Description: Leaves 5–12, 5–30 cm (2–12″) long, narrowly lance-shaped, somewhat erect, green. **Stem** 25–80 or up to 100 cm (10–31 or up to 39″) tall, smooth, green. **Inflorescence** about ⅓ of the total stem, usually 8–25 cm (3–10″) long, with many (up to 75 or more) densely to loosely arranged flowers. **Flowers** about 14–18 or more mm (½–⁷⁄₁₀″ or more) tall and wide, whitish green to pale or bright yellowish green, moderately fragrant. Dorsal **sepal** about 5–7 mm (⅕–¼″) long, egg-shaped; lateral sepals about 6–8 mm (¼–³⁄₁₀″)

long, lance-shaped, spreading outward; **petals** about 6–8 mm (¼–³⁄₁₀″) long, sickle-shaped; dorsal sepal and petals grow forward forming a sheltering hood over the lip, all sepals and petals pale to bright yellowish green; **lip** about 7–9 mm (¼–⅓″) long, long and tapering from a slightly rounded and dilated base (proportionately narrower than the lip of *P. hyperborea*), the long tip curves upward and is often hooked into the "hood" when the flower first opens, later it becomes pendent to slightly recurved—the typical flowering stalk exhibits both extremes and all gradations between, whitish green; the **spur** is slender and tapering towards the tip.

Habitat: highly variable, from open moist or wet areas to shaded woodlands, occasionally occurring in drier soils.

Flowering season: June through August.

Range: *Platanthera huronensis* is found in the Great Lakes region east to Massachusetts and Nova Scotia. It is found throughout the northeast except for southeastern New York, Rhode Island, and eastern Massachusetts.

Comments: As this species is more closely studied, refinements in its physical description and reported range may be made.

Platanthera flava (Linnaeus) Lindley var. **herbiola** (R. Brown) Luer

Etymology: *flava* = "yellow," referring to the yellowish green of the flowers; *herbiola* = "little plant."

Common name: tubercled rein-orchid.

Description: Leaves 2–5, 10–20 cm (4–nearly 8″) long, lance-shaped, green. **Stem** 30–50 cm (12–20″) tall, smooth, green. **Inflorescence** about ⅓–⅖ of the stem length, 10–20 cm (4–8″) long, with 10–30 or up to 40 loosely to densely arranged flowers. **Flowers** about 6 mm (¼″) tall and wide (this figure would be larger if the lip and lateral sepals were not so strongly recurved), yellowish green, each flower growing from the base of a floral bract which on the lower inflorescence greatly exceeds the combined length of the flower and its ovary/stalk. **Sepals** and **petals** similar, about 4 mm (⅙″) long, egg-shaped, sepals green, petals yellowish green; dorsal sepal and petals grow forward to form a sheltering hood over the lip, lateral sepals are recurved; **lip** about 6 mm (¼″) long, oblong with two small triangular lobes near the base, with a small tubercle (fleshy growth) near the center whose shape may vary from wart-like to like a shark fin, tip tongue-shaped and strongly recurved, yellowish green to dull yellow; **spur** about 6 mm (¼″) long and slender.

Habitat: typically found in moist to wet and frequently shaded situations, but large populations will persist in areas that long ago filled in or dried out.

Flowering season: third week of June to first week of August (third to fourth week of July).

Range: *Platanthera flava* as a species is found over most of the United States east of the Mississippi River drainage except south Florida. Variety *herbiola* is found from western North Carolina to Missouri and northward into southern Canada. It is found throughout the northeast.

Comments: After pollination, the ovaries of this species swell quickly, well before the flower wilts. At this time each yellowish-green *P. flava* flower appears to be growing from an oval jade vase.

Platanthera clavellata (Michaux) Luer

Etymology: *clavellata* = "like a small club," referring to the club-like thickening at the tip of the spur.

Common names: little club-spur orchid, small green wood orchid, green rein-orchid.

Description: Leaf single, or rarely two-leaved, 5–15 cm (2–6″) long, lance-shaped, positioned low on the stem, green. **Stem** 10–35 cm (4–nearly 14″) tall, smooth, with conspicuous vertical veining, multiangled if viewed in cross section, with a few bracts, green. **Inflorescence** about ⅕ of the total stem, 13–50 mm (½–2″) long, with 3–15 loosely to densely arranged flowers. **Flowers** about 5 mm (⅕″) tall and wide, all parts yellowish green to pale greenish white, the older flowers in each cluster are twisted about 90° so they seem to be laying on their side. **Sepals and petals** similar, about 4 mm (⅙″) long, the lateral sepals slightly larger, egg-shaped, all parts folded forward, the dorsal sepal and the petals forming a sheltering hood over the lip; **lip** about 6 mm (¼″) long, nearly rectangular but slightly wider at the tip, tip shallowly 3-lobed; **spur** about 10 mm (⅖″) long and thickest near its tip.

Habitat: highly variable, but usually in wet situations ranging from sphagnum fens to riverbanks and roadside ditches.

Flowering season: second week of July to third week of August (third week of July to first week of August).

Range: from Newfoundland to southern Georgia, westward around the Great Lakes to eastern Texas. It is found throughout the northeast.

Comments: The little club-spur orchid is a fairly common wildflower of bogs and other wet situations. It is unique among the *Platantheras* in frequently having its flowers twisted about 90° on their ovary-stalks so that they appear to lay on their sides. I have friends who find the recumbent flowers reminiscent of tiny craneflies or dragonflies, and after observing this orchid, many agree that this image is plausible. Presumably, this unusual angling assists natural pollinators in their work. The little club-spur orchid is also capable of self-pollination, ensuring future generations of "insect-orchids" to delight us all.

Platanthera obtusata (Banks ex Pursh) Lindley

Etymology: *obtusata* = "blunt," referring to the rounded tip of the single leaf.

Common names: blunt-leaf orchid, small northern bog orchid, one-leaf rein-orchid.

Description: Leaf solitary, 5–13 up to 15 cm (2–5 up to nearly 6″) long, egg-shaped with a blunt tip, growing from the base of the stalk. **Stem** 7.5–25 or up to 35 cm (3–10 up to nearly 14″) tall, smooth, 4-angled (nearly square) if viewed in cross-section, green. **Inflorescence** about ⅖ of the total stem, 2.5–10 or up to 17 cm (1–4 or up to nearly 7″) long, with 3–15 loosely arranged flowers. **Flowers** up to 12 mm (⅖″) tall and wide, all parts greenish-white. Dorsal **sepal** about 3 mm (⅛″) long and almost as wide, broadly egg-shaped to nearly triangular; lateral sepals about 5 mm (⅕″) long, lance-like; **petals** about 4 mm (⅙″) long, wide at the lower half but abruptly narrowing above; dorsal sepal and petals grow forward forming a sheltering hood over the lip, the lateral sepals spread outward and are usually recurved; **lip** 6–7 mm (¼″) or more long, narrow and tapering, pendent or recurved; **spur** about as long or longer than the lip.

Habitat: damp forests, wooded fens, and swamps. In the northernmost limits of its range it is found on the exposed and treeless arctic tundra.

Flowering season: fourth week of June to first week of August (first to third week of July).

Range: Circumpolar. From Newfoundland and Nova Scotia westward across most of mainland Canada and Alaska, southward to New York in the east and Colorado in the west. It is also found in northernmost Europe and western Russia. In the northeast it is found in Maine, New Hampshire, Vermont, western Massachusetts and northeastern New York.

Comments: At a glance, this species could be confused with *P. clavellata,* but that species has a broadly tipped lip and a club-like thickening on the point of the spur. In addition, the narrow tips of the petals of this species sometimes point upward, giving the "face" of *P. obtusa* a somewhat "horned" appearance.

Malaxis

Etymology: *Malaxis* = "soft or delicate."

The little malaxises are some of our smallest orchids and often some of the hardest to locate. There are over two hundred *Malaxis* species scattered around the globe, with most of them occurring in Asia and Oceania. Eight species are known from the United States and Canada, and of these three occur in the northeast.

It is appropriate that one of our *Malaxis* species, *M. unifolia,* has a name that translates as "one-leaved," for all three of our species share this trait. These leaves are not fully grown at flowering time and continue to enlarge following fertilization. The common names of green and white malaxis are not very useful in identifying these flowers, for in reality they are all varying shades of green. Despite their diminutive size, the flowering clusters and their tiny lips offer us the needed means of separating the three species.

Key to Malaxis *
(continued from page 104)

I • Flower pedicels very short, flowers appear to spring directly from the stem.
  • Flowers whitish green.
  • Lip with a single thin and tapering tip.
  M. brachypoda → page 165.

II. Flower stalks longer, supporting flowers away from the stem and creating a funnel-shaped or cylindrical inflorescence. Flowers green to yellowish green. Lip with two prominent triangular tips (and a tiny third tip between).
  A. • Inflorescence often funnel-shaped.
     • Flower pedicels usually 5–10mm (1/5″) long.
     • Basal lobes of the lip rounded, shallowly heart-shaped.
     M. unifolia → page 167. *(key continues on p. 164)*

B. • Inflorescence more cylindrical.
   • Flower pedicels up to 5 mm (⅕″) long.
   • Basal lobes elongated, somewhat horn-shaped.
   M. bayardii → page 169.

* This key incorporates information found in *Systematics of Malaxis bayardii and M. unifolia* by P. M. Catling 1991.

Courtesy George H. Beatty

Malaxis brachypoda (Gray) Fernald

Etymology: *brachypoda* = "short pediceled."

Common name: white malaxis.

Description: Leaf solitary, 2.5–7.5 up to 10 cm (1–3 up to 4″) long and usually at least half as wide, broadly egg-shaped, the base enclosing and sheathing the stalk, leaf positioned on the lower portion of the stem, pale green. **Stem** 7.5–15 up to 20 cm (3–6, up to nearly 8″) tall, upper portions angled if viewed in cross section, pale green. **Inflorescence** ½ and often more of the total stem length, usually 2.5–7.5 cm (1–3″) long, with 15–35 or up to 50 loosely arranged flowers. **Flowers** minute, about 4 mm (⅙″) tall and 3 mm (⅛″) wide, all parts greenish white to yellowish green. **Sepals** 2–2.5 mm (1/12–1/10″) long, lance-shaped, dorsal sepal upright, lateral sepals recurved; **petals** 2 mm (1/12″) long, slender and almost thread-like, recurved; **lip** 2–2.5 mm (1/12–1/10″) long, somewhat heart-shaped and triangular, 3-lobed, lateral lobes rounded and folded forward, central lobe triangular, downcurved with the tip recurved.

Habitat: moist and shaded woodlands, often near flowing water.

Flowering season: third week of June to third week of July, a few into late August (fourth week of June to second week of July).

Range: This species is found from Newfoundland south to Pennsylvania and west to British Columbia and southern Alaska. There are also isolated populations in Illinois, Colorado, and California. It is found throughout the northeast except for eastern Massachusetts and Connecticut, Rhode Island, and Long Island.

Comments: Until fairly recently this species was classified as *M. monophyllos* ("single leaf") var. *brachypoda*. The typical form of *M. monophyllos,* is found in Eurasia. Flowers on these plants are twisted around on their stalks so that the lip is uppermost. The central sepal occupies the lowest position on the flower. This unusual flowering orientation may be seen in this hemisphere on var. *diphyllos,* found on the Aleutian Islands. This variety is also unique in having two, or even three, leaves.

Courtesy George H. Beatty

Malaxis unifolia Michaux

Etymology: *unifolia: uni* = "one," *folia* = "leaf."

Common names: green malaxis, green adder's mouth.

Description: Leaf solitary, 2.5–9 cm (1–3½″) long, and often more than half as wide, broadly egg-shaped, the base enclosing and sheathing the lower stem, leaf positioned about half way between the base of the stem and the base of the inflorescence, green. **Stem** 5–30 up to 40 or more cm (2–12 up to 16″ or more) tall, upper portions somewhat angled if viewed in cross section, green. **Inflorescence** about ¼–⅓ of the total stem, usually 2.5–7.5 cm (1–3″) long, with up to 50 flowers in a somewhat funnel-shaped, flat topped-cluster. **Flowers** minute, 4–5 mm (1/6–1/5″) long from the tip of the central sepal to the tips of the lip lobes, 2–4 mm (1/12–1/6″) wide, all parts green to yellowish green, growing at first erect and later drooping on 5–10 mm (1/5–2/5″) long, thin stalks. **Sepals** 1.5–3 mm (1/15–1/8″) long, lance-shaped, central sepal erect, lateral sepals recurved, **petals** up to 2 mm ($^1/_{12}''$) long, slender and almost thread-like, recurved; **lip** 3–4 mm ($^1/_8$–$^1/_6''$) long, broad, heart-shaped at the base and 3-lobed at the tip, the three lobes triangular, the lateral lobes much larger than the tiny central lobe.

Habitat: usually damp to moist woodlands, swamps, roadsides, meadows, and fens.

Flowering season: Second week of June to second week of August, a few into September (second week of June to first week of August).

Range: from Newfoundland to Manitoba, south to Florida and eastern Texas, including all of the northeast. It is also found in Cuba, Mexico, and other parts of Central America.

Comments: It is difficult to describe the flowers of this species in terms of their orientation, for the lip points upright as the flower firsts opens but the entire flower soon twists to the opposite position. The lip has two prominent lobes, which with a little imagination can be seen as the fangs of this green vegetative viper.

Malaxis bayardii Fernald

Etymology: *bayardii*, in honor of Bayard Long (1885–1969).

Common name: Bayard's malaxis.

Description: For most parts of the description, component parts of *Malaxis bayardii* and *M. unifolia* are interchangeable. Measurements for *M. bayardii* seem to mirror those for an average-sized *M. unifolia*. I have a suspicion that, as additional naturalists learn to distinguish between the two species, a number of somewhat larger *M. bayardii* may be recorded. The following details may be useful in identifying *M. bayardii*. **Inflorescence** an extended cylindrical, somewhat flat-topped cluster. **Flowers** growing at first erect and later drooping on thin stalks up to 5 mm (⅕" or less) long. **Lip** with two prominent somewhat horn-like lobes at the base (as opposed to the shallower and rounded heart-shaped base of *M. unifolia.*) Catling (1991) gives a mathematically based key for differentiating the two species by comparing size ratios of the basal and apical (tip) lobes in his definitive paper on *bayardii*.

Habitat: sandy or rocky soil in dry woodlands such as pine-oak barrens and shale barrens.

Flowering season: fourth week of June to second week of August (fourth week of June to second week of August).

Range: *Malaxis bayardii* has been identified from stations in North Carolina, Virginia, Pennsylvania, southeastern New York and Long Island, Connecticut, Rhode Island, and eastern Massachusetts. As more naturalists become aware of this species, it may be discovered in additional areas.

Comments: Bayard's malaxis is a little known orchid which is very difficult to distinguish from *M. unifolia*. Flowers of *M. bayardii* last longer than those of *M. unifolia* before wilting, a trait resulting in *M. bayardii*'s more elongated and cylindrical inflorescence.

Listera:
The True Twayblades

Etymology: *Listera:* in honor of Dr. Martin Lister (1638–1712), an English botanist.

Whenever the word *twayblade* is to be employed, it is a long established tradition to begin by distinguishing the true twayblades, the genera *Listera,* from the so-called false twayblades, the genera *Liparis.* Both share the common name twayblade, meaning two leaved. Despite this similarity, it takes only a glance to tell the two apart, for in *Liparis* the two leaves are found at the very base of the stem, while in *Listera* they grow from a point about one-third to halfway up the stem. The seed capsules provide yet another distinction, those of *Liparis* remain permanently erect, while capsules of *Listera* are drooping when ripe.

There are just over two dozen known species of *Listera* scattered around the globe, most living in temperate to downright cold climates. Seven species are known to occur naturally in the United States and Canada, and one Eurasian species, *Listera ovata,* has somehow found its way to a new home in Ontario. It has been my observation that the different species of *Listera* do not usually share the same immediate habitat, but discoveries of a *L. convallarioides/auriculata* hybrid known as *L.* × *veltmanii* found in Michigan and Maine suggests some "socializing" occurs now and then.

Four species of *Listera* can be found in the northeast. If we compare these looking for similarities, we can create a good description of the genus as it appears worldwide. *Listera* are small orchids, often only a few inches tall and only rarely reaching the ten-inch mark. All have a pair of broad leaves that lack stalks and typically grow opposite from one another at a point about halfway up the stem. The stem has a few sheathing bracts near its base and becomes somewhat pubescent above the leaves. The flowers are usually fairly numerous, and loosely arranged on the upper portion of the stem. Petals and sepals tend to be proportionately small, the lip unusually large. Flowering is followed by a series of drooping seed capsules.

The genus *Listera* was named for Dr. Martin Lister, who lived from 1638–1712. Little is known about the extent of Lister's botanical knowledge, but he did exchange correspondence with the well-known English botanist John Ray. The fact that Ray named an orchid after Lister suggests that he held him in some regard. *Listera* also attracted the attention of some of the other great classical botanists, such as Hooker and Darwin. Darwin is known to have employed the services of his son to stand watch over selected plants so he could report their reproductive habits to his famous father.

The reproductive strategies of *Listera* in some ways appear to be the herbal equivalent of an old-fashioned shotgun wedding. To my mind this image is quite appropriate, for the combination of residing in remote habitats and sporting long-bearded faces (having lengthy lips to the less imaginative) has led me to refer to the twayblades as the hillbillies of the orchid family. The forced wedding is effected by means of a vegetative shotgun, in this case the column, which in the twayblades actually protrudes like the barrel of a gun. When the unsuspecting groom (pollinating insect) alights on the oversized lip to sample the sweet nectar (homebrew?), the sensitive column fires a blast of fast-drying adhesive onto the face of the guest. This glue binds the pollen masses to the insect which then flies off to later deposit his reproductive package on other flowers. One interesting phenomenon of *Listera* is that the flowers have a reputation for persisting for some time after they have carried out their reproductive responsibilities, much to the appreciation and delight of orchid lovers everywhere. It has also been noted that when the time comes to "give birth," the seeds are expelled from the capsule by forceful action. The *Listera* may be little, but it certainly demonstrates an explosive and hair trigger personality!

It is the proportionately oversized lip that is usually employed to differentiate between *Listera* species, and a close examination of those lips is certainly the best guide to separating the four species of true twayblades found in the northeast.

Key to Listera *

(continued from page 104)

I. Lip long and narrow, split at the tip into narrow lobes.
   A. Lip about 10 mm ($^2\!/_5''$) long and four to eight times as long as the petals.
      L. australis → page 174.
   B. Lip 3–5 mm ($^1\!/_8$–$^1\!/_5''$) long and twice as long as the petals, with a horn-like growth at the base.
      L. cordata → page 176.
II. Lip broadly wedge-shaped, with two somewhat rounded lobes at the tip.
   A. Lip 10–12 mm (about $^2\!/_5''$) long, with two auricles (basal lobes) at the base.
      L. auriculata → page 178.
   B. Lip about 10 mm ($^2\!/_5''$) long, abruptly constricted at the base.
      L. convallarioides → page 180.

* Key adapted from Gibson 1905.

Listera australis Lindley

Etymology: *australis* = "southern."

Common names: southern twayblade, shining twayblade.

Description: Leaves 2, 17–25 or up to 35 mm (²/₃–1 or up to 1²/₅″) long, egg-shaped, opposite, growing from a point about halfway up the stem, glossy green. **Stem** 8–20 cm (3–nearly 8″) tall, with a few leaf-like sheaths near the base, pubescent especially above the leaves, purplish green. **Inflorescence** ¼–⅓ of the stem, usually 5–10 cm (2–4″) long, with 5–15 or up to 25 loosely arranged flowers. **Flowers** about 12 mm (nearly ½″) long from the tip of the dorsal sepal to the tips of the lip, 4 mm (⅙″) wide. **Sepals** about 1.5 mm (nearly ¹/₁₂″) long, broadly egg-shaped, spreading, greenish purple to greenish; **petals** about 1.5 mm (nearly ¹/₁₂″) long, oblong to tongue-shaped, spreading outward but recurved at the tips; greenish purple to greenish; **lip** about 10 mm (²/₅″) long, long and thin, the lower third split into slender lobes, usually with a tiny triangular projection at the base of the split, greenish purple to greenish, greener along the center from the base to the split.

Habitat: cool sphagnum fens or red maple swamps.

Flowering season: fourth week of May to second week of July (third week of June to second week of July).

Range: The southern twayblade has a long but discontinuous range. In the northeast it is found over much of New York into western Vermont. It is also found in southern Quebec and Ontario and New Jersey. After skipping several states it is found in coastal lowlands from North Carolina to the tip of Florida and along the Florida Gulf coast into eastern Texas. There is also an isolated population in Tennessee.

Comments: A very rarely observed orchid, in part because of its small size and inconspicuous nature. Like the striped coral-root, "if you haven't seen it in the sunlight, you haven't seen it," because it takes strong sunlight to illuminate the jewel-like quality of its purplish colorations. Because of this orchid's small size, the above advice could easily be altered to "if you haven't seen it in sunlight and through a powerful hand lens, you haven't seen it."

Listera cordata (Linnaeus) R. Brown

Etymology: *cordata* = "heart-shaped," referring to the leaves.

Common name: heart-leaved twayblade.

Description: Leaves 2, 1–2 cm (²/₅–⁴/₅″) or reported up to 4 cm (1³/₅″) long, egg-shaped and typically somewhat heart-shaped at the base, opposite, growing from a point about halfway up the stem, glossy green. **Stem** 7.5–20 or up to 25 cm (3–8 or up to nearly 10″) tall, with a few leaf-like sheaths near the base, smooth on the lower portions to slightly pubescent above, green to purplish green. **Inflorescence** about ¼ of the stem, usually 14–50 mm (½–2″) long, with 4–20 or up to 25 loosely to densely arranged flowers. **Flowers** about 9 mm (⅓″) from the tip of the dorsal sepal to the tips of the lip, 5 mm (⅕″) wide. **Sepals** and **petals** similar, 2–3 mm (¹/₁₂–⅛″) long, egg-shaped or somewhat narrower, spreading, the tips of the petals often recurved, purplish to yellowish green; **lip** 3–6 mm (⅛–¼″) long, long and narrow, the lower half split into slender lobes, with a horizontal horn-like growth near the base, purplish to yellowish green.

Habitat: usually found in cool sphagnum fens and tamarack swamps.

Flowering season: fourth week of May to first week of August (first week of June to first week of July).

Range: around the world in a broad continuous band from Japan through Asia, Europe, Iceland, Greenland, Canada, and Alaska. In the United states it is found around most of the Great Lakes, throughout the northeast and in mountainous areas down to North Carolina. It is also found in the northwest and south to New Mexico in mountainous areas.

Comments: As was true for *L. australis,* strong sunlight brings out the best in the purplish colorations. As the common name suggests, the leaves of *L. cordata* are somewhat heart-shaped, but be careful of relying too heavily on this as the primary identifying feature, for the leaves of *L. autralis* are also occasionally slightly heart-shaped. If the orchids in question are in bloom, any doubts about identification can be quickly erased, for the flowers are easily differentiated. A larger and greener variety, var. *nephrophylla,* has been described from the western United States.

Listera auriculata Wiegand

Etymology: *auriculata* = referring to the auricles, lobes present on the base of the lip.

Common name: auricled twayblade.

Description: Leaves 2, 2.5–5 or up to 6 cm (1–2 or up to 2⅖″) long, somewhat egg-shaped, opposite, growing from a point about halfway up the stem, glossy green. **Stem** 7.5–20 or up to 25 cm (3–8 or up to nearly 10″) tall, with a few leaf-like sheaths near the base, densely pubescent on the upper portions, pale silvery green. **Inflorescence** about ½ of the stem, usually 2.5–7.5 cm (1–3″) long, with 10–20 loosely arranged flowers. **Flowers** about 12–14 mm (⅖–½″) long from the tip of the dorsal sepal to the tips of the lip, 5 mm (⅕″) wide. Dorsal **sepal** about 3.5 mm (⅐″) long, egg-shaped; lateral sepals about 4 mm (⅙″) long, sickle-shaped; **petals** about 3 mm (⅛″) long, sickle-shaped; sepals and petals silvery white and recurved; **lip** 10–12 mm (⅖″) long, oblong, somewhat constricted near the center and divided into 2 oblong lobes near the tip, the base likewise divided into 2 rounded lobe-like auricles (basal lobes), pubescent, silvery-white, greenish along the center.

Habitat: wooded swamps, the banks of streams and rivers.

Flowering season: fourth week of June to third week of July (fourth week of June to first week of July).

Range: from Newfoundland through southern Quebec and Ontario south around Lake Superior and into the very northeastern United States. In our area it is found through most of Maine, in northern New Hampshire, and central Vermont. Recently discovered populations on the banks of the Hudson River in New York's Adirondack Mountains may represent the southern limit of this orchid's range.

Comments: *Listera auriculata* is considered to be the southern counterpart to *L. borealis,* a more northerly orchid preferring even colder habitats. With a little imagination we can visualize the silvery white of the flowers as the touch of frost from those colder climes.

Listera convallarioides (Swartz) Nuttall ex Elliott

Etymology: *convallarioides* = "like a *Convalaria*" (lily-of-the-valley), referring to an imagined similarity between the leaves of these two species.

Common names: broad-leaved twayblade, broad-lipped twayblade.

Description: Leaves 2, 38–50 or up to 70 mm (1½–2, up to 2 ¾″) long, broadly egg-shaped, opposite or nearly so, growing from a point about halfway up the stem, glossy green. **Stem** 10–25 or up to 35 cm (4–10 or up to nearly 14″) tall, with a few leaf-like sheaths near the base, pubescent on the upper portions, silvery green. **Inflorescence** ⅓ to ½ of the stem, usually 38–75 mm (1½–3″) long, with 3–12 or up to 20 loosely arranged flowers. **Flowers** about 12–15 mm (⅖–⅗″) long from the tip of the dorsal sepal to the tips of the lip, 8 mm (³⁄₁₀″) wide, all parts yellowish-green to silvery pale green. **Sepals** and **petals** similar, about 5 mm (⅕″) long, lance-shaped to somewhat sickle-shaped, all parts recurved; **lip** about 8–10 mm (³⁄₁₀–²⁄₅″) long, broadly wedge-shaped, abruptly constricted near the base, tip with 2 broad rounded lobes and a tiny triangular growth between them, pubescent, silvery pale green, green along the center.

Habitat: moist woodlands to the edges of shaded fens and swamps.

Flowering season: third week of June to fourth week of July, a few into August (the third week of June to the second week of July).

Range: from Newfoundland and Nova Scotia west through southern British Columbia. In the western United States it is found in coastal areas south to mid-California and in the Rocky mountains south to Colorado. In the northeast it is found through most of Maine, northern New Hampshire and Vermont, and northeastern New York.

Comments: Of the four discussed *Listera* twayblades, *L. convallarioides* is the most likely to form large colonies.

Liparis
The "False" Twayblades

Etymology: *Liparis* = "greasy," referring to the shiny leaf surface.

Two groups of orchid share the common name of twayblade (two-leaf), *Listera* and *Liparis*. Since historically *Listera* holds the earliest claim to this name, some authors of old referred to *Liparis* as the false twayblades.

Worldwide, *Liparis* is a large genus, with about two hundred and fifty members. Taxonomically, *Liparis* are closely related to the *Malaxis* group and were even included under that name in the early 1800s. Surprisingly only four of these can be found in the United States: the two northeastern species described here, a single species from Florida and another from Hawaii, where in 1987 I had the sad duty of watching the Kileaua lava flows bury several populations.

The two northeastern twayblades are easily differentiated by examining the lips. Those of *L. lilifolia* are pale purple and about 10 mm long, as long or longer than the thread-like petals. The lips of *L. loeselii* are yellowish green and about 5 mm long, somewhat shorter than the thread-like petals. Additionally, it should be noted that *L. lilifolia* typically sports a greater number of flowers than its smaller relation.

Key to Liparis
(continued from page 104)

I. Lip about 10 mm ($2/5''$) long, purplish.
   L. lilifolia → page 183.
II. Lip about 5 mm ($1/5''$) long, yellowish green.
   L. loeselii → page 185.

Liparis lilifolia (Linnaeus) L. C. Richard ex Ker

Etymology: *lilifolia: lili,* referring to the lily family, such as the common trout lily; *folia* = "leaf."

Common names: lily-leaved twayblade, large twayblade, mauve sleekwort.

Description: Leaves 2, 5–13 or up to 18 cm (2–5 or up to 7″) long and half as wide, broadly egg-shaped, bases sheathing the stem, basal, glossy green. **Stem** 10–25 cm (4–nearly 10″) tall, smooth, angled, green. **Inflorescence** $1/2$–$3/4$ of the stem, 5–15 cm (2–6″) long, with 10–20 or up to 27 loosely arranged flowers. **Flowers** about 20 mm ($4/5''$) long from the tip of the dorsal sepal to the tip of the lip, and 12 mm ($2/5''$) wide, mauve to pale purple, growing straight out from the stem. **Sepals** similar, about 10 mm ($2/5''$) long, very narrowly lance-shaped, the dorsal sepal recurved, the lateral sepals curve directly forward under (and sometimes partially supporting) the lip, green; **petals** 5–8 mm ($1/5$–$3/10''$) long, thread-like, growing downward and back toward the stalk, purple; **lip** about 10 mm (²⁄₅″) long, egg-shaped, oriented parallel to the ground and slightly downcurved at the tip, mauve to pale purple often with darker purplish veining.

Habitat: moist woods and thickets, often along ravines or stream banks.

Flowering season: first week of June to second week of July (third week of June to first week of July).

Range: from Massachusetts to North Carolina west to Wisconsin and Arkansas. In the northeast it is or was found in most of New York south of the Adirondack mountains (but is now believed extirpated from Long Island and southeastern New York), in southern Vermont and western Massachusetts, south through Connecticut.

Comments: When compared to the flowers of *L. Loeselii*, *L. lilifolia* indeed earns its common name of "large twayblade," for the darker flowers of *L. lilifolia* appear much larger than those of its paler relation. Unfortunately, they are now also much rarer due to habitat destruction, and the average orchid lover may have to search for many years before finally finding this elusive flower. In contrast to its rarity in the northeast, it is possibly the most common woodland orchid in the midwest.

Liparis loeselii (Linnaeus) L. C. Richard

Etymology: *loeselii,* in honor of Johann Loesel, a German botanist and contemporary of Carl Linnaeus, the father of modern taxonomy, who named this species.

Common names: fen orchid, Loesel's twayblade.

Description: Leaves 2, 5–15 or up to 19 cm (2–6 or up to 7½″) long and one-third as wide, oblong, basal, with bases sheathing the stem, glossy green. **Stem** 5–25 cm (2–nearly 10″) tall, smooth, angled, green. **Inflorescence** ⅓–⅔ of the stem, usually 2.5–10 cm (1–4″) tall, with 2–12 loosely arranged flowers. **Flowers** about 10 mm (⅖″) tall and 5–10 mm (⅕– ⅖″) wide, all parts yellowish green, growing erect from the stem. **Sepals** similar, about 5 mm (⅕″) long, very narrowly lance-shaped and slender, the dorsal sepal growing upward and slightly back toward the stalk, the lateral sepals extend forwards under the lip; **petals** about 5–6 mm (⅕–¼″) long, thread-like, spreading outward; **lip** about 5 mm (⅕″) long, oblong, growing outward then downcurved past the center.

Habitat: in damp to moist areas such as fen meadows and roadside ditches.

Flowering season: second week of June to fourth week of July (fourth week of June to second week of July).

Range: from Nova Scotia to Virginia west to Lake Manitoba and Iowa. Loesel's twayblade is found throughout the northeast except for Cape Cod. It is also found in central Europe.

Comments: *Liparis loeselii* is probably the most commonly encountered twayblade in the northeast, due no doubt to its affection for a wide variety of moist habitats. In spite of this, *L. loeselii* is also one of the most overlooked orchids because of its small size and inconspicuous coloration.

Glossary

Bibliography

Index

Glossary

angiosperm: a flowering plant.

anther: male portion of a flower containing the pollen. In most orchids it is usually divided into two small, rounded packets (pollinia), which are imaginatively referred to as the eyes of the "face" orchids.

axil: the upper angle at the juncture of the stem with a leaf.

basal: growing from the base. This may refer to the base of the stem or the base of flower.

bog: although this name is frequently applied to any wet, somewhat acidic area rich in sphagnum moss, technically a bog is a highly acidic, nutrient poor peatland that relies on atmospheric sources for its moisture. Many of the locations commonly referred to as bogs in the northeast are actually fens.

bract: a small leaf-like structure.

calcareous: containing limestone (calcium carbonate).

capsule: a dried mature ovary. In orchids the capsule contains thousands of dust-sized seeds.

column: a unique reproductive organ that is the major criteria for inclusion in the *Orchidaceae*. The column is made up of the combined male and female sexual organs of the flower.

coniferous: a needle-bearing nonflowering tree, such as the pines, spruces, fir, hemlock, or tamarack.

corm: swollen, underground stem.

deciduous: a broad-leaved flowering tree, such as the maples, elms, or willows.

dicotyledon: a subdivision of flowering plants that produce two cotyledons (seed-leaves) on newly sprouted plants.

disjunct: discontinuous. Used here to refer to species whose major population groups are separated by great distances.

dorsal: upper, as in the uppermost sepal on an orchid flower.

fen: a peatland that receives much of its moisture from ground water sources. When compared to a true bog, a fen is usually less acidic and richer in nutrients and thus richer in the variety of plant life, including orchids, that it supports. Fens are excellent locations to search for many orchid species, true bogs are often orchid poor.

genera: the plural of genus.

genus: a taxonomic classification grouping closely related species together.

globular: in the shape of a globe; round or spherical.

humus: decayed and usually organically rich plant matter found on the forest floor.

inflorescence: that portion of the stem containing the flowers.

lateral: side, referring to sepals and lobes.

lip: the central petal of an orchid, usually but not always found in the lowest position on the flower. Within the *Orchidaceae* the lips are wonderfully varied with regard to their size, shapes, colors, and nectarous offerings, all of which play a prominent role in the reproductive strategy for each species.

lumper: a naturalist with a predisposition for grouping apparently similar flowers into a single species. The opposite of a splitter.

monoandrous: having one fertile anther, as in the lady's slippers.

mottling: blotching.

mycorrhizal: pertaining to roots or root-like structures.

nectary: an alternate name for the spur.

nonresupinate: having a lip oriented in the uppermost position on the flower. Rare in orchids.

ovary: that part of a flower where the seeds develop. Orchids have inferior ovaries, ovaries found between the perianth and the stem. In some orchids, such as the whorled pogonias, they are sometimes mistakenly referred to as flower stalks.

parasitic: surviving by taking nutrients from a living host. *Corrallorhiza* are considered to be indirectly parasitic because they obtain these nutrients through a fungal intermediary.

pedicel: the stalk of a single flower.

pendent: having a downwards orientation. Not intended to infer the freely swinging character of a pendulum or an ornamental pendant.

perianth: the combined petals and sepals of a flower, usually the most colorful part of the flower. In the descriptions the floral height and width actually refer only to the perianth. Although technically the ovaries and the spurs are floral organs, their frequently unusual shape and orientation precude their inclusion in these measurements.

petal: a colorful leaf-like structure which is found between the sepals and the column on an orchid. Orchids have three petals, two lateral petals and the lip, which is a highly modified petal.

pistil: the female parts of a flower, including the stigma, style and ovary.

pollinaria: a stalked mass of pollen with a sticky base.

pubescent: covered with tiny hairs.

recurved: curved back towards the stalk or stem.

resupinate: having the lip oriented towards the lowest position on the flower, typical on most but not all orchid flowers.

reticulation: a net-like pattern, found on the leaves of *Goodyera*.

rhizome: an underground stem.

saccate: sack-like, as in the inflated lips of the lady's slippers.

saprophyte: surviving on the nutrients from dead organisms. Saprophytic plants usually live in areas with a rich layer of humus.

sepal: a sometimes colorful petal-like floral part found between the petals and the base of a flower.

splitter: a naturalist who believes in conferring species status on organisms based on apparently minor physical characteristics. The opposite of a lumper.

species: the primary taxonomic designation that differentiates an organism from all others. This term may be correctly used in either the singular or the plural.

species pair: two distinct species that at first glance are not obviously distinct, such as *Platanthera psycodes* and *P. grandifolia*.

spur: an usually long and slender tubular growth from the base of the lip in some species of orchids. Its function is to store the nectar that attracts pollinators.

stalk: a stem-like structure that supports a vegetative organ such as a leaf or flower.

stamen: the male pollen-bearing part of a flower.

staminode: a broad petal-like growth which is actually an infertile stamen, associated with lady's slippers.

stem: the central supporting organ or structure of a herbaceous plant.

stigma: the female part of the plant which initially receives the male pollen.

style: the female portion of a flower that lies between the stigma and the ovary.

subbasal: growing from very near the base of the stem.

symbiotic: a mutually beneficial relationship between two organisms, such as is shared between an orchid and a subterranean fungus.

taxa: the plural of taxon.

taxon: a general term that can refer to any taxonomic classification level.

taxonomist: a person who specializes in studying the classification of organisms.

Bibliography

Ames, O. *Orchidaceae*. Fasc. I, 1905; fasc. IV, 1910; fasc. VII, 1922. Fasc. I: Houghton Mifflin, Boston. Fasc IV, VII: Merrymount Press, Boston.

Baldwin, H. 1884. *The Orchids of New England*. John Wiley, New York.

Bessette, A. E., and W. K. Chapman. 1992. *Plants and Flowers*. Dover, New York.

Blanchan, N. 1916. *Wild Flowers*. Doubleday, Page, New York.

Britton, N., and A. Brown. 1896. *An Illustrated Flora of the Northern United States, Canada, and the British Possessions*. Vol. 1. Scribner's, New York.

Brown, P. M. 1992. *Platanthera pallida* (Orchidaceae), a New Species of Fringed Orchis from Long Island, New York, USA. *Novon* 2 (4): 308–311.

———. 1993. *A Field and Study Guide to the Orchids of New England and New York*. Orchis Press, Jamaica Plain, Mass.

———. 1995. *North American native Orchid Journal*. 1 (2 and 3).

———. Forthcoming. *Wild Orchids of the Northeastern United States: A Field and Study Guide*. Cornell Univ. Press, Ithaca, N.Y.

Case, F. W., Jr. 1964. *Orchids of the Western Great Lakes Region*. Cranbrook Institute of Science, Bloomfield Hills, Mich.

Catling, P. M., and J. E. Cruise. 1974. *Spiranthes casei*, a New Species from Northeastern North America. *Rhodora*, 76 (808): 526–536.

———. 1991. Systematics of *Malaxis bayardii* and *M. unifolia*. *Lindleyana* 6 (1): 3–23.

Correll, D. S. 1950. *Native Orchids of North America*. Chronica Botanica, Waltham, Mass.

Darwin, C. 1904. *Various Contrivances by Which Orchids Are Fertilized by Insects*. John Murray, London.

———. 1862. *On the Fertilization of Orchids by Insects*. Reprint 1979. E. M. Coleman, Stafford, N.Y.

Gibson, W. H. 1901. *Blossom Hosts and Insect Guests.* Newson, New York.

———. 1905. *Our Native Orchids.* Doubleday, Page, New York.

Gleason, H. A. 1952. *The New Britton and Brown Illustrated Flora of the Northeastern United States and Adjacent Canada.* New York Botanical Garden, New York.

Gupton, O. W., and F. C. Swope. 1987. *Wild Orchids of the Middle Atlantic States.* Univ. of Tennessee Press, Knoxville.

Homoya, M. A. 1993. *Orchids of Indiana.* Indiana Univ. Press, Bloomington, Ind.

House, H. D. 1918. *Wild Flowers of New York.* State Museum Memoir 15, Univ. of the State of New York, Albany.

Johnson, C. W. 1985. *Bogs of the Northeast.* Univ. Press of New England, Hanover, N. H.

Kallunki, J. A. 1976. Population Studies in *Goodyera* (Orchidaceae) with Emphasis on the Hyrid Origin of *G. tesselata*. *Brittonia* 28: 53–75.

Lamont, E. E. 1996. Atlas of the Orchids of Long Island, New York. *Bull. Torrey Club* 123 (2): 157–66.

Lamont, E. E. and Beitel, J. M. 1988. Current Status of Orchids on Long Island, New York. *Torreya* 115 (2): 113–121.

Luer, C. A. 1975. *The Native Orchids of the United States and Canada Excluding Florida.* New York Botanical Garden, Ipswich, Eng.

Mathews, F. S. 1902. *Field Book of American Wild Flowers.* Putnam, New York.

Mitchell, R. S. 1986. *A Checklist of New York State Plants.* New York State Museum Bulletin 458, State Univ. of New York, Albany.

Mitchell, R. S., and C. J. Sheviak. 1981. *Rare Plants of New York State.* New York State Museum Bulletin 445, State Univ. of New York, Albany.

Morris, F., and E. A. Eames. 1929. *Our Wild Orchids.* Scribner's, New York.

Niles, G. G. 1904. *Bog-Trotting for Orchids.* Putnam, New York.

Nylander, O. O. 1935. *Our Northern Orchids.* Star-Herald Publication, Maine.

Petrie, W. 1981. *Guide to Orchids of North America.* Hancock House, Blaine, Wash.

Reddoch, A. H., and J. M. Reddoch. 1993. The Species Pair *Platanthera orbiculata* and *P. macrophylla* (Orchidaceae): Taxonomy, Morphology, Distributions and Habitats. *Lindlyana* 8: 171–187.

Sheviak, C. J. 1982. *Biosystematic Study of the Spiranthes cernua Complex.* New York State Museum, Bulletin 448, State Univ. of New York, Albany.

———. 1994a. *Cypripedium parviflorum* Salisb. Part 1: The Small-flowered Varieties. *American Orchid Society Bulletin* 63(6) : 664–669.

———. 1994b. A New Look at the Taxonomy of Our Yellow Lady's Slippers. *NYFA Newsletter* 5 (2): 3–4.

———. 1995. *Cypripedium parviflorum* Salisb. Part 2: The Larger-flowered Plants and Patterns of Variation. *American Orchid Society Bulletin* 64(6): 606–612.

Stroutamire, W. P. 1974. Relationships of the Purple-fringed Orchids *Platanthera psycodes* and *P. grandiflora*. *Brittonia* 26: 42–58.

Summerhayes, V. S. 1951. *Wild Orchids of Britain.* Collins, London.

Wiegand, K. M. , and A. J. Eames. 1926. *The Flora of the Cayuga Lake Basin,* New York. Cornell Univ., Ithaca.

Williams, J. G., and A. E. Williams. 1983. *Field Guide to Orchids of North America.* Universe Books, New York.

Index

Genera, Species, and Common Names

Adam-and-Eve, 91
adder's mouth, 44
  green, 167
Amerorchis rotundifolia, 110
Aplectrum hyemale, 91
arethusa, 41
Arethusa bulbosa, 41

Beck's tresses, 71
bog candle, 151

calopogon, 38
Calopogon tuberosus, 38
calypso, 33
Calypso bulbosa, 33
Cleistes divaricata, 45
Coeloglossum viride, 116
Corallorhiza, 94
  maculata, 96
  odontorhiza, 102
  striata, 100
  trifida, 98
  wisteriana, 97
coralroot
  autumn, 102
  early, 98
  large, 96
  Macrea's, 100
  northern, 98
  small, 102
  spotted, 96
  striped, 100

cristata, pale, 137
Cypripedium, 12
  acaule, 20
  × *andrewsii* nm. *andrewsii,* 18
  × *andrewsii* nm. *favillianum,* 18
  arietinum, 15
  calceolus, 22
  candidum, 17
  parviflorum var. *makasin,* 28
  parviflorum var. *parviflorum,* 26
  parviflorum var. *pubescens,* 24
  reginae, 30

dragon's mouth, 41

Epipactis helleborine, 113

fairy slipper, 33

Galearis spectabilis, 108
Goodyera, 54
  descipiens, 65
  menziesii, 65
  oblongifolia, 64
  pubescens, 62
  repens, 60
  tesselata, 58

grass-pink, 38
Gyrostachys, 66

Habenaria, 118
hellebore, 113

Isotria, 48
　medeoloides, 51
　verticillata, 49

ladies' tresses
　autumn, 83
　Case's, 77
　hooded, 85
　little, 71
　nodding, 83
　shining, 79
　slender, 73
　spring, 75
　wide-leaved, 79
　yellow nodding, 81
lady's slipper
　large yellow, 24
　pink, 20
　of the Queen, 30
　Queen's, 30
　ram's head, 15
　showy, 30
　small white, 17
　small yellow, 26, 28
　stemless, 20
　white, 17
　yellow, 22, 24
little pearl-twist, 71
little round-leaf, 110
Liparis, 182
　lilifolia, 183
　loeselii, 185

Listera, 171
　auriculata, 178
　australis, 174
　borealis, 179
　convallarioides, 180
　cordata, 176
　　× *veltmanii*, 171
Lysias macrophylla, 148

malaxis
　Bayard's, 169
　green, 167
　white, 165
Malaxis, 163
　bayardii, 169
　brachypoda, 165
　monophyllos, 166
　unifolia, 167
mauve sleekwort, 183
moccasin flower, pink, 20
moss nymph, 41

northern vernalis, 77
nymph, moss, 41

orchid
　blunt-leaf, 161
　broad-leaved, 113
　butterfly, 123
　cranefly, 88
　crested fringed, 135
　crested yellow, 135
　early purple fringed, 125
　fen, 185
　fragrant, 151
　frog, 116
　Goldie's round-leaved, 146

greater purple fringed, 125
green fringed, 129
green rein, 159
helleborine, 113
Hooker's, 142
Hooker's rein, 142
Hooker's round-leaved rein, 142
large butterfly, 125
large purple fringed, 125
large round-leaved, 146
lesser purple fringed, 123
little club-spur, 159
long-bracted, 116
one-leaf rein, 161
pale fringed, 137
prairie fringed, 127
prairie white fringed, 127
ragged fringed, 129
satyr, 116
small green wood, 159
small northern bog, 161
small purple fringed, 123
tall green bog, 153, 155
tall northern green, 153, 155
tall white bog, 151
three-birds, 105
tubercled rein, 157
white fringed, 131
yellow fringed, 133
orchis
  one-leaf, 110
  purple-hooded, 108
  showy, 108
  small round-leaved, 110
Orchis, 109, 111

pale cristata, 137
pearl-twist, little, 71
pink, wild, 41
Piperia, 118

Platanthera, 118
  × *andrewsii*, 139
  × *bicolor*, 139
  blephariglottis, 131
  × *canbyi*, 139
  × *channellii*, 139
  ciliaris, 133
  clavellata, 159
  cristata, 135
  dilatata, 151
  flava, 157
  grandiflora, 125
  hookeri, 142
  huronensis, 155
  hyperborea, 153
  lacera, 129
  leucophaea, 127
  macrophylla, 144
  × *media*, 149
  obtusata, 161
  orbiculata, 144
  pallida, 137
  psycodes, 123
pogonia
  large whorled, 49
  nodding, 105
  rose, 44
  small whorled, 51
  spreading, 45
  whorled, 49
Pogonia ophioglossoides, 44
putty-root, 91

rattlesnake plantain
  downy, 62
  lesser, 60
  Loddiges', 58
  Menzies', 64
  tesselated, 58
round-leaf, little, 110

Serapias, 115
sleekwort, mauve, 183
snake mouth, 44
Spiranthes, 66
  beckii, 72
  casei, 77
  cernua, 83
  × *intermedia,* 67
  lacera, 73
  lucida, 79
  ochroleuca, 81
  praecox, 67
  romanzoffiana, 85
  tuberosa, 71
  vernalis, 75

three-birds orchid, 105
Tipularia discolor, 88
  unifolia, 90

tresses, Beck's, 71
Triphora trianthophora, 105
twayblade
  auricled, 178
  broad-leaved, 180
  broad-lipped, 180
  false, 182
  heart-leaved, 176
  large, 183
  lily-leaved, 183
  Loesel's, 185
  shining, 174
  southern, 174
  true, 171

vernalis, northern, 77

wild pink, 41
winter-leaf, 91